ChaseDream GMAT 备考系列丛书

GMAT 定量推理

数学满分精讲

毕　出◎编著

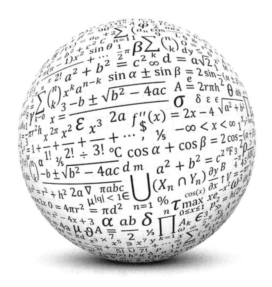

机械工业出版社

CHINA MACHINE PRESS

本书囊括了 GMAT 考试定量推理（即数学）部分的全部知识点及解题技巧。全书共五章。第一章介绍了 GMAT 定量推理部分的题型和解法。第二章是算数，涉及质数、余数、统计和集合等问题。第三章是代数，涉及方程、不等式、数列和函数等问题。第四章是几何，涉及三角形、四边形、圆柱体和解析几何等问题。第五章是文字问题，涉及速率、混合、利率和概率等问题。作者在书中揭示了 GMAT 数学部分的考查重点，具有强大的可推广性和适应性，力求做到内容专而不广，叙述简洁，通俗易懂。

本书适用于所有已经参加或者准备参加 GMAT 考试的考生，也适用于喜好研究 GMAT 考试的同仁。

图书在版编目（CIP）数据

GMAT 定量推理：数学满分精讲：汉文、英文／毕出编著. —北京：机械工业出版社，2022.4

（ChaseDream GMAT 备考系列丛书）

ISBN 978－7－111－70660－1

Ⅰ.①G…　Ⅱ.①毕…　Ⅲ.①高等数学—研究生—入学考试—自学参考资料—汉、英　Ⅳ.①O13

中国版本图书馆 CIP 数据核字（2022）第 071176 号

机械工业出版社（北京市百万庄大街 22 号　邮政编码 100037）
策划编辑：苏筛琴　　　责任编辑：苏筛琴
责任校对：张晓娟　　　责任印制：张　博
中教科（保定）印刷股份有限公司印刷

2022 年 8 月第 1 版·第 1 次印刷
184mm×260mm · 19.75 印张 · 1 插页 · 363 千字
标准书号：ISBN 978－7－111－70660－1
定价：68.00 元

电话服务　　　　　　　　　　网络服务

客服电话：010－88361066　　　机　工　官　网：www.cmpbook.com
　　　　　010－88379833　　　机　工　官　博：weibo.com/cmp1952
　　　　　010－68326294　　　金　书　网：www.golden-book.com
封底无防伪标均为盗版　　机工教育服务网：www.cmpedu.com

特别鸣谢

香港大学 MBA 项目对本书提供的帮助。

Special thanks to
The MBA Programmes, The University of Hong Kong
for their kind help.

Company Visit: Microsoft

Career Internship Fair

The University of Hong Kong — HKU MBA
Learn Business Where Business Is

First and Foremost

As the oldest tertiary education institution in Hong Kong, over 245,000 alumni have been at the forefront of community life in Hong Kong, providing leadership in government, in commerce and industry, in education, and in the arts, sciences and culture. On the world scene, the University of Hong Kong (HKU) has established a solid reputation as a premier international university and a member of the global family of universities.

Worldwide Recognition

The Economist has ranked the HKU MBA Programme No.1 in Asia for 9 consecutive years. Given the relatively short history of the full-time MBA programme at HKU, the HKU Business School's achievement of such high regional and world rankings is impressive indeed. In addition to *The Economist* rankings, the HKU MBA is consistently ranked amongst the top programmes in the world in other scales. The programme has built one of its most dynamic rankings in a very short period of time. HKU as a whole is ranked No. 22 in the world and No.1 in Hong Kong in Quacquarelli Symonds (QS) World University Rankings 2022. The HKU Full-Time MBA Programme is one of the region's youngest MBA programmes, but has managed to grow in *Financial Times* Global MBA Ranking and reached to No.29 in 2021. The university is consistently ranked one of the top universities in Asia in the *Times* Higher Education (THE) World University Rankings.

1-Year Full-Time MBA Programme

The Full-Time MBA Programme is an intensive 1-year programme with three tracks. All students spend nine months in Hong Kong, with field trips to Mainland China. Then, depending on the track they choose, they spend four months in London or New York or Hong Kong/Shanghai, China.

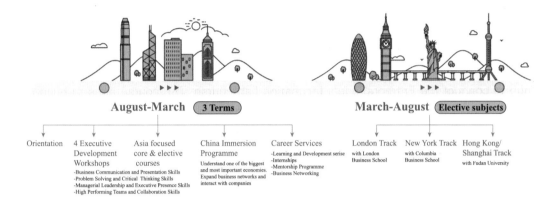

Our Asia-Pacific focus, will give you a distinct advantage in building your career in the region, or indeed anywhere in the world, as Asia now lies at the centre of many business ventures worldwide.

The China Factor

The HKU MBA's dual focus on Asia and China business renders the programme highly relevant to the wider region's dynamic, ever-evolving business environment. The China focus is achieved through special courses with China elements, regional case studies, field trips, seminars and conferences, Chinese language training and company visits.

The London & New York Connection: an Expanded Global Vision

The unique partnerships with world-renowned business schools such as London Business School and Columbia Business School provide our students with unmatched opportunities in terms of educational experience and networking. London and New York are both world-class cities and offer great environments for business education along with international exposure for our students.

Case-based Approach

The HKU MBA Programme adopts an experiential-learning approach, with the extensive use of business cases that enables students to become effective problem-solvers and decision-makers. These cases are written by our own professors and are published by our renowned Asia Case Research Centre (www.acrc.hku.hk).

Mentorship Programme

This programme connects current students to senior professionals to build mentor-mentee relationship. Students can obtain real-life business knowledge from senior executives and top business leaders. The real case sharing and professional insights from experienced mentors offer students an engaging and enriching learning experience. Recent graduates engage with students as buddies in sharing their information about studies, school experience and network in Hong Kong.

Entrepreneurial Incubation Lab

HKU MBA partners with Cyberport Academy, part of Hong Kong's largest start-up incubation hub to conduct the Entrepreneurial Incubation Lab course. The course aims to build up entrepreneurs' mentality that leverage on the training, cultivate the capabilities of start-up founders to identify the market opportunities, and to engage in practical workshops to develop investor decks and present start-up pitch.

Executive Development Workshops

A series of Executive Development Workshops have been created to encourage the students to fully engage in a unique learning experience, while sharpening their soft skills, including Business Communication and Presentation Skills, Problem Solving and Critical Thinking Skills, Managerial Leadership and Executive Presence, and High Performing Teams and Collaboration Skills.

Collaborative & Interactive Culture

The programme's relatively small class sizes along with diverse profiles allow for extensive interaction and collaboration.

China Immersion Programme

As many of our students are preparing for careers in China, this programme will help them to build language capabilities, better understand a different business system and expand networks in the region. The course is composed of company visits, language training, executive talks, alumni sharing sessions and cultural activities. They will be given an eye-opening opportunity to experience the country from different perspectives: from indulging themselves in the cultural richness to experiencing the world's leading technology.

Contacts

Website: https://mba.hkubs.hku.hk
Email: ftmba@hku.hk
Telephone: (852)3962–1241
Room 204, Block B, Cyberport 4, 100 Cyberport Road, Hong Kong

ChaseDream 总编推荐序

具有数学天赋的我们，为什么在 GMAT 数学考试上会经常翻车呢？

从以前"不考 50 + 不是中国人"，到现在越来越多的同学拿着 45 分上下的 ESR 找我们分析原因；从被无视时的"人均 50 +"，到特别重视后的成绩"一落千丈"。过去的 10 年，问题到底出在哪里？是 GMAT 考试对数学的要求提高了？还是大家的数学基础普遍变差了？

与大量考生深入沟通后，我们找到了一些原因：

为了让本来不太重要的 GMAT 数学专项辅导课卖得更好，很多机构一味地强调那些难题和偏题。习题库中甚至常常出现奥数题的影子。同学们听完老师对这些题目"精巧"的分析和讲解后，瞬间"大彻大悟"。GMAT 数学的存在感刷上去了，但离 GMAT 考查的核心却越来越远了。

又回到那个老生常谈的问题：GMAT 考试，到底想要考查我们什么呢？我们又应该如何复习呢？

于是在推出第一本 GMAT 复习书之后的第八年，一直"无视" GMAT 数学的我们，终于还是决定写一本书，告诉大家 GMAT 数学到底在考查什么。

开始吧！

steven

ChaseDream 总编

前　言

数学是很多考生心中的"痛"。因为它不仅意味着需要背诵许多公式，掌握很多技巧，更意味着需要有清晰的数学思维，而数学思维在很多人看来是"天赋"，是后天难以习得的。实则不然，大部分人的数学思维都是可以锻炼的，并且只要方式得当，很快即可熟练掌握。

数学思维和处事思维是一脉相承的，因为它们都需要从两个步骤入手：

（1）定义问题；

（2）拆分问题。

所谓"定义问题"，就是当我们看到一个数学题后，要先根据题目的信息定义出这道题在本质上涉及哪一个或哪几个数学问题。例如：

一共8个杯子，现在它们全部杯底向上。假设我们每次能且仅能翻转其中的3个杯子，那么，最少经过几次翻转能让这8个杯子全部杯口向上？

实际上，这个很"生活化"的问题本质上是一个倍数问题。只有当进行前一次翻转后还剩下3个或3的倍数个杯底向上的杯子时，我们才能完成题目要求，因此，这个问题应定义为倍数问题。由此可以得出第一步思路，即假设进行若干次翻转后，还剩下3个或6个杯底向上的杯子。

所谓"拆分问题"，顾名思义，就是将无法直接解决的问题拆分成若干个简单且可以一步解决的小问题，进而逐步求解。例如，拆分例题中的倍数问题就是单独考虑例题中的每一次翻转，直到剩余的杯底向上的杯子的个数为3或6。

第一次，无论如何只能翻转前3个杯子。

第二次，若想尽量保证杯底向上的杯子是6个，则我们可以这样翻转，即两个杯口向上的杯子翻转为杯底向上，一个杯底向上的杯子翻转为杯口向上。

第三次和第四次分别翻转 3 个杯底向上的杯子即可使 8 个杯子全部杯口向上。综上，题目可求解，即最少需要翻转 4 次。

培养优秀的数学思维就是在不同题目中一次次地重复进行"定义问题"和"拆分问题"。本书的目的是希望借助 GMAT 数学的知识点和题目来帮助考生培养出这种优秀的数学思维。期待大家可以在读完整本书后获得巨大提升。

感谢王钰儿、陈晨和龙甜的帮助，没有你们，本书无法顺利写就。感谢我的妻子郭宁对本书提供的许多建设性意见。

毕 出

于 2022 年春

目 录

第一章

定量推理简介

定量推理是中国考生的强项。GMAT 考试中一共有31 道数学考题，均为五选一的单选题，限时 62 分钟内答完。GMAT 数学考查的知识点范围虽然和中国大陆地区的小学和初中的知识点范围重合，但是，GMAT 对"数学"的考查更注重对题目的逻辑把握，而不是单纯的应用公式和计算。

下面，让我们先来看一道难度"较高"的例题。

A certain experimental mathematics program was tried out in 2 classes in each of 32 elementary schools and involved 37 teachers. Each of the classes had 1 teacher and each of the teachers taught at least 1, but not more than 3, of the classes. If the number of teachers who taught 3 classes is n, then the least and greatest possible values of n, respectively, are

(A) 0 and 13   (B) 0 and 14   (C) 1 and 10

(D) 1 and 9   (E) 2 and 8

无论采取不等式、方程式、排列组合，还是任何"高级"公式，这道例题都不是那么容易求解，反而容易越算越乱。那如何入手呢？现在让我们暂时忘掉"套公式"这个传统数学题目的解法，把重点放在理解数学情景上。

假设现在我的左手边有 64 个班，右手边有 37 名教师。要求是：每个班需要 1 名教师管理，每名教师不能闲着，但也不能太累，最少管理 1 个班，最多管理 3 个班。问：管理 3 个班的教师最多和最少分别是多少？

先讨论最少的情况。实际上，最少有几名教师管理 3 个班，意思无非就是问，有没有教师是不得不受累带 3 个班的。如果我们给右手边的 37 位教师中的每一位平均分配 2 个班级，则他们带的总班级数量已经超过 64 个。换句话说，可以实现没有任何一名教师是不得不带 3 个班的。因此，带 3 个班的教师数量最少肯定为 0。

最多的情况会有点复杂。最直白的一个办法是，把 64 个班全部拆成 3 个班一组，能拆出 21 个完整组，这就构成了 3 个班最多的情况。但有个致命的问题，即这样的拆分，虽是能保证有"最多的 3 个班"，但显然会有教师"没事干"，不能满足题干的要求。为了解决这个问题，想保证教师们都有事情干，我们可以先从 64 个班里拆出 37 个班，先给每名教师一人分配一个班，这样会剩下 27 个待分配的班。因为每名教师已经带了 1 个班，现在只需把剩下的 27 个班拆成 2 个班一组，再分配给教师们就可以实现管理 3 个班的教师最多。显然，最多能拆出 13 组来分配给 13 名教师，还剩一个单独的班随意给一名教师。因此，最多可以让 13 名教师带三个班。

综上，答案为 A。

例题的解法中并没有用到任意一条公式。许多数学考题都有类似的现象，即考查的重点不在于记忆和背诵公式，而更多是在于理解题意，用逻辑而不是纯粹的计算来解题。

1.1 ▸ 两种题型

数学部分有两种题型——Problem Solving 和 Data Sufficiency。两者均为 5 选 1 的考题，并且题量各占一半。PS 题型与初中数学考试中的选择题几乎没有区别。

例题 1

If x percent of 40 is y, then $10x$ equals

(A) $4y$　　　　　(B) $10y$　　　　　(C) $25y$

(D) $100y$　　　(E) $400y$

解题方法很简单：先计算出答案，然后选择和计算结果一致的选项。

由例题的条件可知，

因为 $40 \times \left(\dfrac{x}{100}\right) = y$；

所以 $x = 5\dfrac{y}{2}$，

则 $10x = 25y$，

因此，答案为选项 C。

而 DS 考题比较新颖，请先看下面的例题。

例题 2

Is $x = 1$?

(1) $(x+1)(x-1) = 0$

(2) $(x+2)(x-2) = 0$

(A) Statement (1) ALONE is sufficient, but statement (2) alone is not sufficient.

(B) Statement (2) ALONE is sufficient, but statement (1) alone is not sufficient.

(C) BOTH statements TOGETHER are sufficient, but NEITHER statement ALONE is sufficient.

(D) EACH statement ALONE is sufficient.

(E) Statements (1) and (2) TOGETHER are NOT sufficient.

我们可以观察到，仅凭题干是无法求解的，而且在题干下给出了两个条件。而所有 DS 考题的五个选项都是一模一样的，把它们翻译为中文则是：

(A) 条件 1 单独是充分的，条件 2 单独是不充分的。

(B) 条件 2 单独是充分的，条件 1 单独是不充分的。

(C) 两个条件加一起是充分的，任何一个条件单独都是不充分的。

(D) 每个条件单独都是充分的。

(E) 条件 1 和条件 2 相加都是不充分的。

可见，这种题型不要求我们算出具体数字和答案，而是要求我们推断出哪个条件能得出题干的解。因此，我们也不一定需要算出具体答案，只需判断是否可以求解的充分性即可。

这里的充分判断具体表现在两个层面上 —— 唯一性和充分性。

先来谈唯一性。例题的条件 1，通过解方程可以得到 $x = 1$ 或 $x = -1$。这个结果看起来是可以解出 x 的数值的，但是，题干问 "x 是否等于 1"。也就是说，基于条件 1，我们不能确定 x 究竟应该等于 1 还是等于 -1，所以我们不能认为条件 1 是充分的。这点就被称为"唯一性"，即只有当一个条件能推出问题的唯一解时，它才是充分的。

再来谈充分性。例题的条件2，通过解方程可以得到 $x=2$ 或 $x=-2$。正是因为无论 x 等于哪个值，我们均能确定 x 必然不等于1，所以我们认为条件2是充分的。这点就被称为充分性，即当一个条件能确定回答题干的问题时，它就是充分的。

综上，例题2的答案为选项 B。

正是因为 DS 题有上述特点，所以它考查我们的不是计算能力，而是"寻找最简需求"的能力。

最简需求的定义是：充分解决问题所需的最少条件。

为了能确定题目的最简需求，我们应该使用的解题方法如下：

首先，只看题干，不看两个条件，优先确定题干的最简需求是什么。

其次，检查哪个条件（或两个条件加在一起）能满足该需求。

下面我们用几个例题来练习一下。

例题 3

Machines X and Y produced identical bottles at different constant rates. Machine X, operating alone for 4 hours, filled part of a production lot; then Machine Y, operating alone for 3 hours, filled the rest of this lot. How many hours would it have taken Machine X operating alone to fill the entire production lot?

(1) Machine X produced 30 bottles per minute.

(2) Machine X produced twice as many bottles in 4 hours as Machine Y produced in 3 hours.

(A) Statement (1) ALONE is sufficient, but statement (2) alone is not sufficient.

(B) Statement (2) ALONE is sufficient, but statement (1) alone is not sufficient.

(C) BOTH statements TOGETHER are sufficient, but NEITHER statement ALONE is sufficient.

(D) EACH statement ALONE is sufficient.

(E) Statements (1) and (2) TOGETHER are NOT sufficient.

首先，根据题干可知，设机器 X 和机器 Y 的速度分别为 x 和 y，则总产量应为： $4x+3y$。若由机器 X 独立完成，则需要的时间为：

$$(4x+3y)x,$$

化简可得:

$$4+3\frac{y}{x},$$

由此可知，若想知道机器 X 单独工作的时间，至少需要知道 y 和 x 的比值，即机器 X 的速度和机器 Y 的速度的比值。

其次，我们要检查哪个条件能给出"机器 X 的速度和机器 Y 的速度的比值"。

条件 1 给出了机器 X 的速度，无法推出两个机器的速度比，因此条件 1 不充分。

条件 2 给出了机器 X 和机器 Y 在不同时间下产量的比值。因为产量除以时间等于速度，所以条件 2 相当于告诉了我们两个机器的速度比。因此，条件 2 单独是充分的。

综上，答案为 B。

例题 4

If $xy>0$, does $(x-1)(y-1)=1$?

(1) $x+y=xy$

(2) $x=y$

(A) Statement (1) ALONE is sufficient, but statement (2) alone is not sufficient.

(B) Statement (2) ALONE is sufficient, but statement (1) alone is not sufficient.

(C) BOTH statements TOGETHER are sufficient, but NEITHER statement ALONE is sufficient.

(D) EACH statement ALONE is sufficient.

(E) Statements (1) and (2) TOGETHER are NOT sufficient.

解:

首先，根据题干得出: $(x-1)(y-1)=xy-x-y+1$。问题是这个式子是否等于 1。显然，我们至少需要知道 xy, x 和 y 的关系。

其次，我们要检查哪个条件能给出这三者的合理关系。

条件 1 给出了 $x + y = xy$，显然此时 $xy - x - y = 0$，即 $(x-1)(y-1) = xy - x - y + 1 = 1$，故条件 1 充分。

条件 2 给出了 $x = y$，因为并不能得出 xy，x 和 y 三者的关系，所以条件 2 不充分。

综上，答案为 A。

例题 5

Is x^2 greater than x?

(1) x^2 is greater than 1.

(2) x is greater than -1.

(A) Statement (1) ALONE is sufficient, but statement (2) alone is not sufficient.

(B) Statement (2) ALONE is sufficient, but statement (1) alone is not sufficient.

(C) BOTH statements TOGETHER are sufficient, but NEITHER statement ALONE is sufficient.

(D) EACH statement ALONE is sufficient.

(E) Statements (1) and (2) TOGETHER are NOT sufficient.

解:

首先根据题干得出，x^2 在 $x > 1$ 的情况下必然大于 x；在 $1 > x > 0$ 的情况下 x^2 小于 x；在 $0 > x$ 的情况下，因为 x^2 是正数，x 为负数，此时 x^2 必然大于 x。

其次，我们要检查哪个条件能给出 x 的取值范围。

条件 1 可以解出 $x > 1$ 或 $x < -1$。显然在这两个范围内，x^2 均大于 x，因此条件 1 充分。

条件 2 给出 $x > -1$，无法确定 x 是否满足 $1 > x > 0$，因此条件 2 不充分。

综上，答案为 A。

1.2 ▸ 数学的第一性原理

很多同学之所以小学数学"没学好",乃至中学、大学的数学变成"弱项",并不是因为数学天赋差,而是从根本上就没能理解数学知识的底层逻辑。那么,现在就让我们一起从源头开始"重塑"数学,相信你会爱上它的。

所有的数学命题最终都应归结为关于自然数的命题,这一点是现代数学的指导原则。正如数学家克隆尼克(L. Kronecker)说过:"上帝创造了自然数,其余的是人的工作"。

我们需要把自然数及其两种基本运算——加法和乘法——当作已知概念(对,你没看错,不是四则运算,这也是小学的误区之一),它们无法被证明,是数学的第一性原理。

什么是加法呢?例如:

$$○○ + ○○○ = ○○○○○。$$

加法的定义是,把两组小球都堆到一起。该例子中,把两个小球和三个小球堆到一起构成了五个小球。

什么是乘法呢?例如:

$$○○ × ○○○ = \begin{array}{c}○○\\○○\\○○\end{array}。$$

乘法的定义是,将其中一个数当成行,另一个数当成列,例如将 2 当成行,将 3 当成列,则有:

$$○○ × ○○○ = \begin{array}{c}○○\\○\\○\end{array}。$$

之后用小球将空余部分补全,得到最后的答案是:

$$○○ × ○○○ = \begin{array}{c}○○\\○○\\○○\end{array}。$$

显然,也可以将 3 当成行,2 当成列,结果不变(乘法的交换定律)。

很多我们熟悉的公式均来自于这个乘法的定义。例如，在计算长方形面积时，其公式为：

长方形面积 = 长 × 宽。

那为什么把求面积的公式定义为"长 × 宽"呢？请看下图：

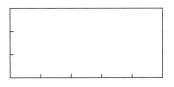

我们把一个格子的长度定义为 1，叫"单位长度"。由此可知，该长方形的长为 5 个单位长度，宽为 3 个单位长度。如果我们把每个单位长度看作一个小球，则有：

如果让小球铺满整个长方形，则有：

○ ○ ○ ○ ○

○ ○ ○ ○ ○

○ ○ ○ ○ ○

因此，如果想知道整个长方形能容纳多少个小球，那么根据乘法定义，只需把长和宽相乘即可。

任何规则物体的面积和体积均是利用乘法规则定义的。不但是几何图形，等差数列的求和公式也是这个道理。

等差数列的求和公式和梯形面积公式是一样的：

$$S_n = \frac{n(a_1 + a_n)}{2},$$

都可以简记为：上底加下底的和，乘高，除以 2。等差数列形如：

如果把四周连起来，则有：

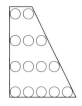

可以发现，上图是一个非常标准的直角梯形。只不过，直角梯形的质地非常密（密到不可见），等差数列的质地是松散的。但基于乘法定义，让小球铺满整个图形的"面积"公式都是相同的。

基于加法和乘法这两种运算，我们可以推理出它们的逆运算——减法和除法。加法和乘法在自然数的体系中是永远有效的，即无论取多少个自然数做加法或乘法，所得的结果必然依旧在自然数体系内。

但它们的逆运算，减法或除法，则没有这个必然性。例如：

$$9 - 3 = 6。$$

6 依然是自然数，但 3 - 9 的运算结果明显不在自然数体系内。

基于此，我们就人为地规定了负数，即当被减数小于减数时，减法运算的结果就为负数。

又例如：

$$\frac{9}{3} = 3。$$

3 依然是自然数，但 $\frac{9}{4}$ 的结果就不在自然数体系内了，基于此，我们就规定了余数的概念；$\frac{4}{9}$ 显然也不在自然数体系内，基于此，我们就规定了分数的概念。

当然，从另一个角度，根据自然数和乘法又可以推理出质数、因子等概念。

由此，数字因运算的需要而被拓展为自然数、整数、负数和分数等。看起来它们已经足以度量所有的长度了。例如，我们规定"—"为单位长度（即为1）。任意一段线段，我们都可以用 n 个 "—" 来表示。如果待测线段的长度不是完整的单位，那么我们可以利用分数把单位长度的线段进行无限细分。因此，在常理上，我们总是可以利用单位长度来测量出"待测线段"的长度。

但是，基于非常简单的运算和证明，我们还可以得到一些"反常理"的数：

$$1 + 1 = x^2 。$$

我们假设，x 可以用分数进行度量，因此设它为 $\dfrac{p}{q}$，同时假设此时 $\dfrac{p}{q}$ 已无法再约分，即最简分数（lowest term）。

$$\left(\frac{p}{q}\right)^2 = 2 ,$$
$$p^2 = 2q^2 。$$

由于两个奇数相乘必定是奇数，所以此时 p 必为偶数（因为 $2q^2$ 必然是偶数）。

我们可以把 p 表示为 $2r$，代入则有：

$$4r^2 = 2q^2 ,$$
$$q^2 = 2r^2 ,$$

同理，此时 q 也必为偶数。

还记得这个推理最开始的假设吗？我们假设，$\dfrac{p}{q}$ 已无法再约分。但两个偶数必然是可以再约分的，这就和假设相矛盾。

换句话说，$x^2 = 2$ 中的 x 无法用整数和分数来表达。

但是，利用直尺和圆规，我们可以很容易地在数轴上画出 x 的位置（半径为 $\sqrt{2}$，以原点为圆心画弧，弧和数轴的交点就是 x 的位置）。

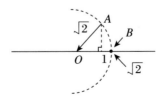

如上图，OA 的长度为 $\sqrt{2}$。以它为半径画弧，交数轴于点 B，点 B 的位置就是 $\sqrt{2}$ 的位置。但就是这个位置，就算我们拼命细分数轴，哪怕精确到无限小（譬如亿万分之一的单位长度），都只能无限趋近却无法准确描述它。

因此，这个在常识上很容易找到且度量的点，用单位长度和分数却永远无法表达。正是因为这种数是如此不合常理，所以它们被称为：无理数。

无理数的出现至今还让许多爱思考的人兴奋不已。

由此这些数字对应的元素可以构成另一个基本概念——集合。如同数字可以通过加法和乘法形成另外的数字一样，一些集合通过某些运算可以形成另外的集合。这就是集合代数的由来。交集、并集、补集、容斥原理等，这些概念是结合数字算法和常识后的产物。

当集合内的元素都是数字时，研究集合内元素的规律，可以得到"数列"的概念。等差、等比、递推都是集合中数字元素的关系。

集合代数不仅能研究数字，还可以用于研究非数字概念。非数字概念中，最重要的莫过于"事件"了。如果把集合代数用于研究事件，则形成了另一门学科——概率论。

这也是为什么在计算独立事件的概率时，我们用的不是普通的加减法，而是集合论中的"容斥原理"（因为事件是以集合代数为根基推理得出的，而非以自然数为根基）。

利用这些可度量（有理数）或不可度量（无理数）的数为单位，我们可以构造出在生活中常见的几何图形。例如：三角形、圆形、正方形、长方形、梯形、正方体、长方体、棱锥等。将基本几何图形放在坐标系中，形成了解析几何。解析几何需要作图描点工具，这个工具可以被定义为"函数"。函数的一种特殊情况，即因变量为常数时，就得到了方程的概念。用方程来建模实际生活中的事件，就有了混合问题、工程问题和追击问题等。

由此可见，无论初等数学还是高等数学，均可以由自然数、加法和乘法这三个简单到极致的概念推理得出。

由此，这些数学问题基本上可以分为 4 大类——算数、代数、几何和文字问题。下面，我们就按照这 4 大类问题分类来系统讲解 GMAT 数学的问题。

第二章

算 数

提到算数，首当其冲的就是数论了。数论说容易很容易，说难也真的很难。说它容易，是因为基本上所有数论的知识性内容，我们在小学课本中都已经学过了；而说它难，是因为它一直是数学家们研究的重要课题，而且变化多端。

数论，顾名思义，就是研究数字的理论。数字又包括了整数、分数、小数等。如果曾经参加过小学的奥林匹克数学竞赛，那么你一定记过一些神奇的十进制数字规律。例如，9 乘 2~9 中任意数字后，可以得到一个两位数结果，这个结果的两位数字相加必然等于 9，例如：

$$3 \times 9 = 27, \ 2 + 7 = 9;$$
$$5 \times 9 = 45, \ 4 + 5 = 9。$$

类似这样的规律几乎是无穷无尽的。但是，GMAT 考试的重点和小学奥数完全不同，GMAT 的重点在于利用数学概念考查考生的推理能力，只需把握数字的核心概念即可，那些神奇的规律均不会刻意涉及。

2.1 ▸ 整数的正负性和奇偶性

2.1.1 ▸ 正负性

整数（integer）包括正整数（positive integer）、负整数（negative integer）和零（zero）。

例如：2 是正整数；−2 是负整数。

正数和负数的四则运算经常考查到。请记住在乘除运算中，"正负得负，负负得正"，例如：

$$(-2) \times 2 = -4; \ (-2) \times (-2) = 4。$$

例题 1

Which of the following represent positive numbers?

Ⅰ. $-3-(-5)$

Ⅱ. $(-3)(-5)$

Ⅲ. $-5-(-3)$

(A) Ⅰ only　　　(B) Ⅱ only　　　(C) Ⅲ only

(D) Ⅰ and Ⅱ　　(E) Ⅱ and Ⅲ

解：

$$-3-(-5)=-3+5=2,$$
$$(-3)(-5)=15,$$
$$-5-(-3)=-5+3=-2,$$

综上，答案为 D。

例题 2

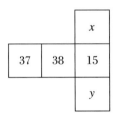

In the figure above, the sum of the three numbers in the horizontal row equals the product of the three numbers in the vertical column. What is the value of xy?

(A) 6　　　　(B) 15　　　　(C) 35

(D) 75　　　　(E) 90

解：

GMAT 考试特别喜欢用一些复杂的框架来考一些十分简单的问题。例如这道例题，

实际上就是考查 $\dfrac{37+38+15}{15}=6$，答案为 A。

What is the sum of 3 consecutive integers?

(1) The sum of the 3 integers is less than the greatest of the 3 integers.

(2) Of the 3 integers, the ratio of the least to the greatest is 3.

(A) Statement (1) ALONE is sufficient, but statement (2) alone is not sufficient.

(B) Statement (2) ALONE is sufficient, but statement (1) alone is not sufficient.

(C) BOTH statements TOGETHER are sufficient, but NEITHER statement ALONE is sufficient.

(D) EACH statement ALONE is sufficient.

(E) Statements (1) and (2) TOGETHER are NOT sufficient.

解：

条件 1 说，这三个数的和小于这三数中最大的数字。这个条件只能证明，这三个数中有负整数，但无法确定这三个数的和是多少。故条件 1 不充分。

条件 2 说，这三个数中，最小的数和最大的数的比值是 3。由此可知，这三个数必然是 −3，−2 和 −1。请注意，1，2，3 不符合条件，因为 1:3 是 $\frac{1}{3}$ 而不是 3。故条件 2 充分。

综上，答案为 B。

例题 4

Is x an integer?

(1) x^2 is an integer.

(2) $\frac{x}{2}$ is not an integer.

(A) Statement (1) ALONE is sufficient, but statement (2) alone is not sufficient.

(B) Statement (2) ALONE is sufficient, but statement (1) alone is not sufficient.

(C) BOTH statements TOGETHER are sufficient, but NEITHER statement ALONE is sufficient.

(D) EACH statement ALONE is sufficient.

(E) Statements (1) and (2) TOGETHER are NOT sufficient.

解：

题目问的是，x 是否是整数。

条件 1 说，x^2 是整数。显然，一个数的平方是整数，不代表这个数是整数。比如$\sqrt{3}$的平方是 3，但显然$\sqrt{3}$不是整数。故条件 1 不充分。

条件 2 说，$\dfrac{x}{2}$不是整数。所有偶数除以 2 后都是整数，但所有奇数除以 2 后都不是整数，非整数除以 2 也都是非整数。故条件 2 不充分。

两个条件同时成立时，依然无法确定 x 是否是整数。

综上，答案为 E。

例题 5

A certain clock marks every hour by striking a number of times equal to the hour, and the time required for a stroke is exactly equal to the time interval between strokes. At 6:00 the time lapse between the beginning of the first stroke and the end of the last stroke is 22 seconds. At 12:00, how many seconds elapse between the beginning of the first stroke and the end of the last stroke?

(A) 72　　　　(B) 50　　　　(C) 48　　　　(D) 46　　　　(E) 44

解：

这道题的主要难度在于理解题意。

6 点钟时，敲钟 6 下间隔 5 次，共 11 次。"敲钟和间隔时间一样"告诉我们如果总耗时是 22 秒，则每一次耗时为：

$$22 \div 11 = 2s。$$

因此，12 点钟时，敲 12 下，隔 11 次，共 23 次。

$$23 \times 2 = 46s。$$

综上，答案为 D。

例题 6

Each of the integers from 0 to 9, inclusive, is written on a separate slip of blank paper and the ten slips are dropped into a hat. If the slips are then drawn one at a time without replacement, how many must be drawn to ensure that the numbers on two of the slips drawn will have a sum of 10?

(A) Three.　　(B) Four.　　(C) Five.　　(D) Six.　　(E) Seven.

解：

这道题的核心在于看懂题目的问题。它其实就是在问，$0 \sim 9$ 一共 10 个数字做成标签，最倒霉的情况下，我要拿多少个标签才能保证其中两个相加为 10。显然，如果依次拿到 0，1，2，3，4，5，则这 6 个均没法让其中两个凑成 10。拿到第 7 个标签，无论是几，均能和前面 6 个标签中的某一个凑成 10。因此，答案为 E。

例题 7

×	a	b	c
a	d	e	f
b	e	g	h
c	f	h	j

Each entry in the multiplication table above is an integer that is either positive, negative, or zero. What is the value of a?

(1) $h \neq 0$

(2) $c = f$

(A) Statement (1) ALONE is sufficient, but statement (2) alone is not sufficient.

(B) Statement (2) ALONE is sufficient, but statement (1) alone is not sufficient.

(C) BOTH statements TOGETHER are sufficient, but NEITHER statement ALONE is sufficient.

(D) EACH statement ALONE is sufficient.

(E) Statements (1) and (2) TOGETHER are NOT sufficient.

解：

要想判断 a 的值，势必需要知道一些字母所代表的数字。

条件 1 说，h 不为 0。单纯凭借这个条件，无法得知 a 的值，故条件 1 不充分。

条件 2 说，$c = f$。观察图表可知，$c \times a = f$。如果 $c = f$，那么 a 有两种可能。要么 $a = 1$；要么 $a = c = f = 0$。依然无法确定 a 的值，故条件 2 不充分。

条件 1 + 条件 2，若 h 不等于 0，因为 $c \times b = h$，则证明 c 必然不为 0。结合条件 2，a 必然为 1。故条件 1 + 条件 2 充分。

综上，答案为 C。

例题 8

A collection of 36 cards consists of 4 sets of 9 cards each. The 9 cards in each set are numbered 1 through 9. If one card has been removed from the collection, what is the number on that card?

(1) The units digit of the sum of the numbers on the remaining 35 cards is 6.

(2) The sum of the numbers on the remaining 35 cards is 176.

(A) Statement (1) ALONE is sufficient, but statement (2) alone is not sufficient.

(B) Statement (2) ALONE is sufficient, but statement (1) alone is not sufficient.

(C) BOTH statements TOGETHER are sufficient, but NEITHER statement ALONE is sufficient.

(D) EACH statement ALONE is sufficient.

(E) Statements (1) and (2) TOGETHER are NOT sufficient.

解：

一套卡片上的数字之和为 45，所以 4 套卡片上的数字总和为 180。

条件 1 说，拿出一张卡片后，剩余的 35 张卡片上的数字总和的个位数字为 6。显然，要想让 180 减去一个一位数后个位数是 6，那么这个数只能是 4，故条件 1 充分。

条件 2 说，拿出一张卡片后剩下的 35 张卡片上的数字总和为 176。显然，180 − 176 = 4，拿出的卡片上的数字为 4。故条件 2 充分。

综上，答案为 D。

2.1.2 ▶ 奇偶性

偶数（even number）是所有能被 2 整除（divisible）的数（包括 0，因为 0 也能被 2 整除），可以表达为 $2k$（k 是任意整数）；奇数（odd number）是所有不能被 2 整除的数，可以表达为 $2k+1$（k 是任意整数）。

关于奇偶性，有许多基础运算规则和两条特殊规则。

基础运算规则是奇数和偶数加减乘除的规则，例如：奇数 + 奇数 = 偶数；偶数 × 奇数 = 偶数等。对于这些运算规律，完全不需要逐一背诵，因为那很容易记错。只需拿出几个典型的奇数和偶数一试便知。

例如：偶数取 2；奇数取 3。

  若问：奇数 + 偶数 = ? 则因为 3 + 2 = 5，5 是奇数，所以奇数 + 偶数 = 奇数。

  若问：奇数 × 奇数 + 偶数 = ? 则因为 3 × 3 + 2 = 11，11 是奇数，所以奇数 × 奇数 + 偶数 = 奇数。

例题 9

If x is an odd integer, which of the following is also an odd integer?

(A) $x^2 + 1$ (B) $6x + 4$ (C) $x^3 + 3$ (D) $x^2 + x + 3$ (E) $x^3 + 1$

解：

类似这类考题，虽然考查的是奇偶性，但实则无需用奇偶的运算规则来推算，只需要代入最简单的奇数 1，就可以发现，只有 D 选项的结果依然为奇数 5，因此答案为 D。

例题 10

If the set S consists of five consecutive positive integers, what is the sum of these five integers?

(1) The integer 11 is in S, but 10 is not in S.

(2) The sum of the even integers in S is 26.

(A) Statement (1) ALONE is sufficient, but statement (2) alone is not sufficient.

(B) Statement (2) ALONE is sufficient, but statement (1) alone is not sufficient.

(C) BOTH statements TOGETHER are sufficient, but NEITHER statement ALONE is sufficient.

(D) EACH statement ALONE is sufficient.

(E) Statements (1) and (2) TOGETHER are NOT sufficient.

解：

问题问的是 5 个连续的数字之和是多少。

条件 1 说，11 在集合中，但是 10 不在集合中。这表示，五个连续的数字必然是 11，12，13，14，15。故条件 1 充分。

条件 2 说，S 中所有的偶数之和是 26。那么，5 个连续的数字中，要么包含 3 个偶数，要么包含 2 个偶数。如果包含 3 个偶数，必然首尾是偶数，中间是偶数，如果 $S = \{6, 7, 8, 9, 10\}$，则偶数的和为 24，小于 26。如果 $S = \{8, 9, 10, 11, 12\}$，则偶数的和为 30，大于 26，所以，S 中不可能包含 3 个偶数。

因此，S 只能包含两个偶数，而包含两个相邻偶数且和为 26 的只能为 12 和 14，所以，S 只能为 $\{11, 12, 13, 14, 15\}$。故条件 2 充分。

综上，答案为 D。

例题 11

For each positive integer k, let $a_k = 1 + \dfrac{1}{k+1}$. Is the product a_1, a_2, $\cdots$, a_n an integer?

(1) $n+1$ is a multiple of 3.

(2) n is a multiple of 2.

（A）Statement（1）ALONE is sufficient, but statement（2）alone is not sufficient.

（B）Statement（2）ALONE is sufficient, but statement（1）alone is not sufficient.

（C）BOTH statements TOGETHER are sufficient, but NEITHER statement ALONE is sufficient.

（D）EACH statement ALONE is sufficient.

（E）Statements（1）and（2）TOGETHER are NOT sufficient.

解：

$$a_1 = 1 + \frac{1}{2} = \frac{3}{2}$$

$$a_2 = 1 + \frac{1}{3} = \frac{4}{3}$$

$$a_3 = 1 + \frac{1}{4} = \frac{5}{4}$$

…

…

由此可知，当 n 是奇数时，a_1 乘到 a_n 必然为 $\frac{n+2}{2}$，因为 n 为奇数，所以 $\frac{n+2}{2}$ 必不为整数；当 n 是偶数时，a_1 乘到 a_n 也为 $\frac{n+2}{2}$，因为 n 为偶数，所以 $\frac{n+2}{2}$ 必为整数。

条件 1 说，$n+1$ 是 3 的倍数，这表明 n 除以 3 余 2（关于这一点，可以用高斯同余定理的加法定理来证明，后文会有详细的解读），因此，n 可以为 2，5，8，11 等，不能确定 n 是奇数还是偶数，所以条件 1 不充分。

条件 2 说，n 是 2 的倍数，这表明 n 为偶数，所以 $\frac{n+2}{2}$ 必为整数，故条件 2 充分。

因此，答案为 B。

例题 12

If $x < y < z$ and $y - x > 5$, where x is an even integer and y and z are odd integers, what is the least possible value of $z - x$?

（A）6 （B）7 （C）8 （D）9 （E）10

定量推理简介
第一章

算数
第二章

代数
第三章

几何
第四章

文字问题
第五章

解：

根据题意可知：

$$y > 5 + x。$$

由于 x 是偶数，所以 $5 + x$ 必然为奇数。又因为 y 是奇数，所以 $y \geqslant 5 + x + 2$。也就是说，y 最小为 $5 + x + 2$。因为 $z > y$ 且 z 是奇数，所以同理，z 最小为 $5 + x + 2 + 2$。

综上，$z - x$ 最小值为 $5 + 2 + 2 = 9$。答案为 D。

例题 13

If x and z are integers, is $x + z^2$ odd?

（1）x is odd and z is even.

（2）$x - z$ is odd.

（A）Statement（1）ALONE is sufficient, but statement（2）alone is not sufficient.

（B）Statement（2）ALONE is sufficient, but statement（1）alone is not sufficient.

（C）BOTH statements TOGETHER are sufficient, but NEITHER statement ALONE is sufficient.

（D）EACH statement ALONE is sufficient.

（E）Statements（1）and（2）TOGETHER are NOT sufficient.

解：

条件 1 说，x 是奇数且 z 是偶数。显然，给定了 x 和 z 的奇偶性，必然可以知道 $x + z^2$ 的奇偶性，故条件 1 充分。

条件 2 说，$x - z$ 是奇数。如果 $x - z$ 是奇数，那么 x 和 z 必然一个是奇数一个是偶数。由于 z^2 的奇偶性和 z 必然相同，所以 $x + z^2$ 必然是一个奇数加一个偶数，其和必然是奇数，故条件 2 充分。

综上，答案为 D。

例题 14

If m and n are positive integers, is n even?

(1) $m(m+2)+1=mn$

(2) $m(m+n)$ is odd.

(A) Statement (1) ALONE is sufficient, but statement (2) alone is not sufficient.

(B) Statement (2) ALONE is sufficient, but statement (1) alone is not sufficient.

(C) BOTH statements TOGETHER are sufficient, but NEITHER statement ALONE is sufficient.

(D) EACH statement ALONE is sufficient.

(E) Statements (1) and (2) TOGETHER are NOT sufficient.

解：

根据条件 1，如果 m 是奇数，则 $m(m+2)$ 必为奇数，$m(m+2)+1$ 必为偶数，则此时 n 必为偶数；如果 m 是偶数，则 $m(m+2)+1$ 必为奇数且 mn 必为偶数，此时与假设矛盾，不成立。因此，该条件可以判断 n 必然为偶数，故条件 1 充分。

根据条件 2，如果 m 是奇数，则 n 必为偶数；如果 m 是偶数，则 $m(m+n)$ 永远不可能为奇数，不成立。因此，该条件可以判断 n 必然为偶数，故条件 2 充分。

综上，答案为 D。

例题 15

Set S consists of 20 different positive integers. How many of the integers in S are odd?

(1) 10 of the integers in S are even.

(2) 10 of the integers in S are multiples of 4.

(A) Statement (1) ALONE is sufficient, but statement (2) alone is not sufficient.

(B) Statement (2) ALONE is sufficient, but statement (1) alone is not sufficient.

(C) BOTH statements TOGETHER are sufficient, but NEITHER statement ALONE is sufficient.

（D）EACH statement ALONE is sufficient.

（E）Statements（1）and（2）TOGETHER are NOT sufficient.

解：

条件 1 说，S 中有 10 个数是偶数。正整数中的任何数，要么是偶数，要么是奇数。有 10 个偶数，必然另外 10 个是奇数，故条件 1 充分。

条件 2 说，10 个整数是 4 的倍数。因为偶数是 2 的倍数，所以是 4 的倍数的数肯定是偶数，但无法确定其余的 10 个整数是否是偶数。故条件 2 不充分。

综上，答案为 A。

2.2 ▸ 质数、因数、最大公约数和最小公倍数

质数（prime number）是那些只能被 1 和它本身整除的正整数（1 自己不算质数），例如：2，3，5，7，11 等。其中，2 是唯一的一个偶质数（even prime number）。因为偶数必然能被 2 整除，而质数的定义是不能被除自己和 1 以外的其他数整除，所以除 2 之外的所有质数必然是奇数。其余的数被称为合数（composite number）。请注意，0 和 1 既不是质数也不是合数。

因数（factor）一般是针对某一个数字定义的。例如，12 的因数是 1，2，3，4，6，12。也就是说，所有能整除 12 的数都是 12 的因数。

那些同时也是质数的因数，被称为质因数（prime factor）。例如，12 的质因数为 2 和 3。

GMAT 数学经常会考查一个数的质因数分解（prime factorization）。所谓质因数分解，意思是将一个数不断拆分，直到拆分为所有的因数都是质数为止。最常用的拆分手段是"短除法"，例如：

$$
\begin{array}{r|r}
5 & 150 \\
\hline
5 & 30 \\
\hline
3 & 6 \\
\hline
& 2
\end{array}
$$

因此，$150 = 2 \times 3 \times 5^2$。

例题 1

If x is the product of the positive integers from 1 to 8, inclusive, and if i, k, m, and p are positive integers such that $x = 2^i \, 3^k \, 5^m \, 7^p$, then $i + k + m + p =$

（A）4 　　　　（B）7 　　　　（C）8 　　　　（D）11 　　　　（E）12

解：

$x = 1 \times 2 \times 3 \times 4 \times 5 \times 6 \times 7 \times 8 = 2 \times 3 \times 2^2 \times 5 \times 2 \times 3 \times 7 \times 2^3 = 2^7 \times 3^2 \times 5 \times 7$，

因此，$i = 7$；$k = 2$；$m = 1$；$p = 1$。$i + k + m + p = 11$。答案为 D。

例题 2

If the product of the integers w, x, y, and z is 770, and if $1 < w < x < y < z$, what is the value of $w + z$?

（A）10 　　　　（B）13 　　　　（C）16 　　　　（D）18 　　　　（E）21

解：

分解 770 的质因数可得：

$$770 = 2 \times 5 \times 7 \times 11,$$

由于 $w < x < y < z$，所以 $w = 2$；$z = 11$。

答案为 B。

对质因数分解的考查共有四种形式——最大质因数、整除与倍数、质因数与因数个数和最大公约数与最小公倍数。

2.2.1 ▸ 最大质因数

请注意，GMAT 考试经常会用几种不同的说法表述因数的倍数（multiple），但这不能改变其本质含义，比如以下三句话的意思均相同：

A is a multiple of B.

B is a factor of A.

A is divisible by B.

例如：如果 B 是 $2^2 \times 3^2$，而 A 是 $2^2 \times 3^4 \times 7$，那么我们可以认为：

A 包含了 B，或 A 是 B 的倍数，或 B 是 A 的因数，或 A 能被 B 整除。

又例如：如果 B 是 $2^2 \times 3^2$，而 A 是 $2 \times 3^4 \times 7$，那么我们可以认为：

A 不包含 B，或 A 不是 B 的倍数，或 B 不是 A 的因数，抑或 A 不能被 B 整除。

最大质因数的考法比较直白，会直接问考生某个数的最大质因数是哪个选项。显然，要想知道一个合数的最大质因数，我们需要先对这个数字做质因数分解，然后找到最大的那个质因数。例如：

$$15 = 3 \times 5;$$

15 的最大质因数是 5。

例题 3

What is the greatest prime factor of $2^{100} - 2^{96}$?

(A) 2　　　　(B) 3　　　　(C) 5　　　　(D) 7　　　　(E) 11

解：

求解某个数的最大质因数，就是在求解这个数的质因数分解。由于题干中给的是减法（因为减法和乘法分属于两则基本运算，所以两者之间无法直接转换），而我们求的是质因数乘积，所以需要运算出差值后再进行质因数分解。

$$2^{100} - 2^{96} = 2^{96}(2^4 - 1) = 2^{96} \times 3 \times 5,$$

因此，答案为 C。

例题 4

What is the greatest prime factor of $417 - 228$?

(A) 2　　　　(B) 3　　　　(C) 5　　　　(D) 7　　　　(E) 11

解：

先计算出差值，即

$$417 - 228 = 189,$$

再将 189 进行质因数分解：

$$189 = 3^3 \times 7,$$

因此，答案为 D。

例题 5

For every even positive integer m, $f(m)$ represents the product of all even integers from 2 to m, inclusive. For example, $f(12) = 2 \times 4 \times 6 \times 8 \times 10 \times 12$. What is the greatest prime factor of $f(24)$?

（A） 23   （B） 19   （C） 17   （D） 13   （E） 11

解：

根据题目的描述，$f(24) = 2 \times 4 \times 6 \times 8 \times 10 \times 12 \times 14 \times 16 \times 18 \times 20 \times 22 \times 24$。其中 22 可以被分解为 2×11。11 是这些数字能分解出的最大质因数，所以答案为 E。

例题 6

Let S be the set of all positive integers having at most 4 digits and such that each of the digits is 0 or 1. What is the greatest prime factor of the sum of all the numbers in S?

（A） 11   （B） 19   （C） 37   （D） 59   （E） 101

解：

题目的意思是：集合 S 中包含了所有 4 位数以下的数字，这些数字有个要求，就是每一个数位只能是 0 或者 1。由此可知，本题的解题关键在于要耐心找到所有这样的 4 位数，然后求和再进行质因数分解。

从一位数开始找，这些数为：

1

10, 11

100, 101, 110, 111

1000, 1001, 1010, 1011, 1100, 1101, 1110, 1111

求和的结果为 8888。

$$8888 = 2^3 \times 11 \times 101$$

因此，答案为 E。

2.2.2 ▸ 整除与倍数

整除（divisible）与倍数（multiple）问题本质上也是分解质因数问题。试想，若某个数是 7 的倍数（或能被 7 整除），那么这个数分解质因数后，这些质因数中必定有 7；反之亦然，若一个合数能分解出 7，那么它一定是 7 的倍数，也能被 7 整除。

因此，但凡看到问题中问的是整除或者倍数问题，我们的思路一定是先将问题问的数想办法进行质因数或因数分解，进而最终求解。

例题 7

If the integer n is divisible by 12, which of the following must be true?

(A) $\dfrac{n}{6}$ is an integer.　　　(B) $\dfrac{n}{9}$ is an integer.　　　(C) $\dfrac{n}{15}$ is an integer.

(D) $2\dfrac{n}{9}$ is an integer.　　　(E) $3\dfrac{n}{15}$ is an integer.

解：

如果 n 能被 12 整除，则 $n = 2^2 \times 3 \times k$，即 n 必然至少包含两个 2 和一个 3。选项 A 表示 n 至少包含一个 2 和一个 3，必然正确；选项 B 表示 n 至少包含两个 3，不一定正确；选项 C 表示 n 至少包含一个 3 和一个 5，不一定正确；选项 D 和选项 B 一样；选项 E 表示 n 必然包含至少一个 5，不一定正确。综上，答案为 A。

例题 8

The difference $942 - 249$ is a positive multiple of 7. If a, b, and c are nonzero digits, how many 3-digit numbers abc are possible such that the difference $abc - cba$ is a positive multiple of 7?

(A) 142　　　　(B) 71　　　　(C) 99　　　　(D) 20　　　　(E) 18

解：

首先需要理解问题，这道题问的是 $abc - cba$ 是否是 7 的倍数。因为整除和倍数问题本质上都是质因数分解问题，所以这道题其实就是让我们将 $abc - cba$ 进行质因数分解。

$$abc - cba = 100a + 10b + c - (100c + 10b + a) = 99a - 99c = 99(a - c) = 3 \times 3 \times 11 \times (a - c)$$

显然，如果想让这个数是 7 的倍数，只能让 $a - c$ 是 7 的倍数。

因为题干中给出了 a，b 和 c 都是非零的个位数字，所以 $a - c$ 只能等于 7。

当 $a = 9$；$c = 2$ 时，$a - c$ 等于 7，此时 b 是 1~9 中的任意数，共有 9 种可能。

当 $a = 8$；$c = 1$ 时，$a - c$ 等于 7，此时 b 也是 1~9 中的任意数，共有 9 种可能。

因此，一共有 18 种可能。答案为 E。

例题 9

Is the sum of two integers divisible by 10?

(1) One of the integers is even.

(2) One of the integers is a multiple of 5.

(A) Statement (1) ALONE is sufficient, but statement (2) alone is not sufficient.

(B) Statement (2) ALONE is sufficient, but statement (1) alone is not sufficient.

(C) BOTH statements TOGETHER are sufficient, but NEITHER statement ALONE is sufficient.

(D) EACH statement ALONE is sufficient.

(E) Statements (1) and (2) TOGETHER are NOT sufficient.

解：

题目问的是两个整数的和能否被 10 整除。只有知道这两个整数和的质因数分解结果中是否有 2 和 5，才能知道它们的和是否能被 10 整除。

条件 1 说，其中一个数是偶数。显然，知道其中一个数的情况，完全不能知道两个数相加之后的情况，故条件 1 不充分。

条件 2 说，其中一个数是 5 的倍数。和条件 1 相同，条件 2 也不充分。

条件1+条件2，知道两个数分别的情况，依然无法得知这两个数相加后的情况，故条件1+条件2不充分。

综上，答案为E。

例题 10

If x and y are integers greater than 1, is x a multiple of y?

(1) $3y^2 + 7y = x$.

(2) $x^2 - x$ is a multiple of y.

(A) Statement (1) ALONE is sufficient, but statement (2) alone is not sufficient.

(B) Statement (2) ALONE is sufficient, but statement (1) alone is not sufficient.

(C) BOTH statements TOGETHER are sufficient, but NEITHER statement ALONE is sufficient.

(D) EACH statement ALONE is sufficient.

(E) Statements (1) and (2) TOGETHER are NOT sufficient.

解：

首先理解问题。题目问的是 x 是否为 y 的倍数，其实就是问 x 进行质因数分解后，是否含有 y。

条件1可以写为：

$$y(3y+7) = x,$$

这表明，x 必然是 y 的倍数。故条件1充分。

条件2可以写为：

$x(x-1)$ 是 y 的倍数。

这个条件只能表明 $x(x-1)$ 能分解出 y，但不能确定是 x 能分解出 y，还是 $x-1$ 能分解出 y。故条件2不充分。

综上，答案为A。

例题 11

If n is a positive integer and the product of all the integers from 1 to n, inclusive, is a multiple of 990, what is the least possible value of n?

(A) 10   (B) 11   (C) 12   (D) 13   (E) 14

解：

如果 n 是 990 的倍数，则表示 n 至少应为 $2 \times 3^2 \times 5 \times 11$。由此可知，$n$ 进行质因数分解后至少含有 11，否则无法提供 11 这个质因数，故答案为 B。

例题 12

If p is a positive integer, what is the value of p?

(1) $\dfrac{p}{4}$ is a prime number.

(2) p is divisible by 3.

(A) Statement (1) ALONE is sufficient, but statement (2) alone is not sufficient.

(B) Statement (2) ALONE is sufficient, but statement (1) alone is not sufficient.

(C) BOTH statements TOGETHER are sufficient, but NEITHER statement ALONE is sufficient.

(D) EACH statement ALONE is sufficient.

(E) Statements (1) and (2) TOGETHER are NOT sufficient.

解：

题目问：p 的值是多少？

条件 1 说，$\dfrac{p}{4}$ 是质数，这表明，p 分解质因数为：$2^2 \times$ 某个质数，且这个质数是唯一的一个质因数，但由于无法确定该质数是多少，所以单独通过这个条件无法确定 p 的值。

条件 2 说，p 能被 3 整除，这只能表示 p 分解质因数后含有 3，也单独无法确定 p 的值。

由条件 1 + 条件 2 可知，那个唯一的质因数就是 3，p 分解质因数必然为 $2^2 \times 3$，故答案为 C。

例题 13

Is the integer k divisible by 4?

(1) $8k$ is divisible by 16.　　　　　　(2) $9k$ is divisible by 12.

(A) Statement (1) ALONE is sufficient, but statement (2) alone is not sufficient.

(B) Statement (2) ALONE is sufficient, but statement (1) alone is not sufficient.

(C) BOTH statements TOGETHER are sufficient, but NEITHER statement ALONE is sufficient.

(D) EACH statement ALONE is sufficient.

(E) Statements (1) and (2) TOGETHER are NOT sufficient.

解:

题目问的是, k 是否能被 4 整除, 也就是问, k 进行质因数分解后, 是否会含有两个 2。

条件 1 说, $8k$ 能被 16 整除, 也就是 k 中必然有一个 2, 故条件 1 不充分。

条件 2 说, $9k$ 能被 12 整除, 这证明 k 中必然有两个 2 (因为, 12 进行质因数分解后包含两个 2, 而 9 是不包含 2 的), 故条件 2 充分。

答案为 B。

例题 14

Is the integer r divisible by 3?

(1) r is the product of 4 consecutive positive integers.　　　(2) $r < 25$

(A) Statement (1) ALONE is sufficient, but statement (2) alone is not sufficient.

(B) Statement (2) ALONE is sufficient, but statement (1) alone is not sufficient.

(C) BOTH statements TOGETHER are sufficient, but NEITHER statement ALONE is sufficient.

(D) EACH statement ALONE is sufficient.

(E) Statements (1) and (2) TOGETHER are NOT sufficient.

解:

题目问的是, r 是否能被 3 整除, 也就是问, r 进行质因数分解后, 是否会含有 3。

条件 1 说, r 是 4 个连续的整数乘积。显然, 任意 3 个连续的数都必然有一个数是 3 的倍数, 更别说是 4 个连续的数了, 故条件 1 充分。

条件 2 说, r 小于 25, 小于 25 的整数显然不一定能被 3 整除, 故条件 2 不充分。

综上, 答案为 A。

Is the integer n a multiple of 15 ?

(1) n is a multiple of 20.

(2) $n + 6$ is a multiple of 3.

(A) Statement (1) ALONE is sufficient, but statement (2) alone is not sufficient.

(B) Statement (2) ALONE is sufficient, but statement (1) alone is not sufficient.

(C) BOTH statements TOGETHER are sufficient, but NEITHER statement ALONE is sufficient.

(D) EACH statement ALONE is sufficient.

(E) Statements (1) and (2) TOGETHER are NOT sufficient.

解：

题目问：n 是否是 15 的倍数。这其实是在问：n 进行质因数分解后是否含有 3 和 5。

条件 1 说，n 是 20 的倍数，这表示 n 进行质因数分解后含有两个 2 和一个 5，依然不能确定是否含有 3。故条件 1 不充分。

条件 2 说，$n + 6$ 是 3 的倍数。因为 6 是 3 的倍数，所以若 $n + 6$ 想是 3 的倍数，n 必然也是 3 的倍数（关于这一点的证明，相信大家在学完高斯同余定理后会明白）。但仅通过条件 2，无法得知 n 进行质因数分解后是否含有 5。故条件 2 不充分。

条件 1 + 条件 2 可保证 n 进行质因数分解后既含有 3，又含有 5，因此条件 1 + 条件 2 充分。

综上，答案为 C。

If the integers a and n are greater than 1 and the product of the first 8 positive integers is a multiple of a^n, what is the value of a?

(1) $a^n = 64$

(2) $n = 6$

（A）Statement（1）ALONE is sufficient, but statement（2）alone is not sufficient.

（B）Statement（2）ALONE is sufficient, but statement（1）alone is not sufficient.

（C）BOTH statements TOGETHER are sufficient, but NEITHER statement ALONE is sufficient.

（D）EACH statement ALONE is sufficient.

（E）Statements（1）and（2）TOGETHER are NOT sufficient.

解：

题目说前 8 个正整数相乘是 a^n 的倍数，问 a 是多少。所有的倍数问题都是质因数分解问题，所以这道题其实就是在问 a 的质因数分解模式。

条件 1 说，a^n 等于 2^6。显然此时 a 可以等于 2，也可以等于 4，还可以等于 8。无法确定 a 的具体数值。故条件 1 不充分。

条件 2 说，$n = 6$。$n = 6$ 实际上是告诉我们，前 8 个数相乘，积分解质因数后必须含有 6 个 a。显然此时 a 只能等于 2。因此，条件 2 是充分的。

综上，答案为 B。

例题 17

If n is a multiple of 5 and $n = (p^2)q$, where p and q are prime numbers, which of the following must be a multiple of 25?

（A）p^2 （B）q^2 （C）pq

（D）p^2q^2 （E）p^3q

解：

题目问的是哪个选项一定是 25 的倍数。也就是问，哪个选项的质因数分解模式中一定包含两个 5。条件中告诉了我们 $n = (p^2)q$，且 n 是 5 的倍数。这表明，要么 p 等于 5，要么 q 等于 5。因此，我们要找的选项，无论 p 还是 q 等于 5，都能保证含有两个 5。显然，只有将 p 和 q 都平方，才能保证无论 p 还是 q 等于 5，选项才能含有两个 5。因此，答案为 D。

例题 18

Is the integer n a prime number?

(1) $24 \leq n \leq 28$

(2) n is not divisible by 2 or 3.

(A) Statement (1) ALONE is sufficient, but statement (2) alone is not sufficient.

(B) Statement (2) ALONE is sufficient, but statement (1) alone is not sufficient.

(C) BOTH statements TOGETHER are sufficient, but NEITHER statement ALONE is sufficient.

(D) EACH statement ALONE is sufficient.

(E) Statements (1) and (2) TOGETHER are NOT sufficient.

解：

题目问的是 n 是否是质数。

条件1给出了 n 的范围，显然24, 25, 26, 27, 28 都是合数，从而可以确定 n 必然不是质数，所以条件1是充分的。

条件2说 n 不能被2或者3整除。不能被2或者3整除的数不代表不能被其他数整除，所以不能确定 n 是否是质数，条件2不充分。

综上，答案为 A。

例题 19

$n = 2^4 \cdot 3^2 \cdot 5^2$ and positive integer d is a divisor of n. Is $d > \sqrt{n}$?

(1) d is divisible by 10.

(2) d is divisible by 36.

(A) Statement (1) ALONE is sufficient, but statement (2) alone is not sufficient.

(B) Statement (2) ALONE is sufficient, but statement (1) alone is not sufficient.

(C) BOTH statements TOGETHER are sufficient, but NEITHER statement ALONE is sufficient.

（D）EACH statement ALONE is sufficient.

（E）Statements（1）and（2）TOGETHER are NOT sufficient.

解：

题目问的是 d 是否大于 $\sqrt{n}$，而 $\sqrt{n}$ 的质因数分解模式为 $2^2 \times 3 \times 5$。这道题其实就问 d 的质因数分解模式是什么。

条件 1 说，d 能被 10 整除，这就表示 d 进行质因数分解后至少含有一个 2 和一个 5，故条件 1 不充分。

条件 2 说，d 能被 36 整除，这表示 d 进行质因数分解后至少含有两个 2 和两个 3，故条件 2 不充分。

条件 1 + 条件 2 则保证了 d 进行质因数分解后至少含有两个 2，两个 3 和一个 5，此时一定大于 $2^2 \times 3 \times 5$，故条件 1 + 条件 2 充分。

综上，答案为 C。

还有两条特殊规则可以单独记忆，可以方便快速完成考题。

3 个连续正整数的乘积必然是 6 的倍数。

2 个连续偶数的乘积必然是 8 的倍数。

这两条性质可以用除法的概念加以证明，证明过程下一节"余数"中会细讲，大家现在直接应用即可。

例题 20

If b is the product of three consecutive positive integers c, $c + 1$, and $c + 2$, is b a multiple of 24?

（1）b is a multiple of 8.

（2）c is odd.

（A）Statement（1）ALONE is sufficient, but statement（2）alone is not sufficient.

（B）Statement（2）ALONE is sufficient, but statement（1）alone is not sufficient.

（C）BOTH statements TOGETHER are sufficient, but NEITHER statement ALONE is sufficient.

（D）EACH statement ALONE is sufficient.

（E）Statements（1）and（2）TOGETHER are NOT sufficient.

解:

依然是首先理解问题。题目问 b 是否是 24 的倍数，其实就是问 b 进行质因数分解后，是否含有 $2^3 \times 3$。由于题干中给出了 b 是由三个连续数的乘积构成的，所以 b 必然是 6 的倍数，也就是 b 进行质因数分解后含有 1 个 3 和 1 个 2。

只要条件中能给出 b 进行质因数分解后含有两个 2，即是充分的。

条件 1 说 b 是 8 的倍数，所以表明 b 进行质因数分解后至少含有 3 个 2，故条件 1 充分。

条件 2 说 c 是奇数，只有两个连续的偶数才是 8 的倍数。若 c 是奇数，则 $c+1$ 是偶数，$c+2$ 是奇数。比如 $c=5$，则显然 b 不是 8 的倍数；再比如 $c=7$，则 b 是 8 的倍数。因此，条件 2 是不充分。

综上，答案为 A。

例题 21

If n is a positive integer and r is the remainder when $n^2 - 1$ is divided by 8, what is the value of r ?

（1）n is odd.

（2）n is not divisible by 8.

（A）Statement（1）ALONE is sufficient, but statement（2）alone is not sufficient.

（B）Statement（2）ALONE is sufficient, but statement（1）alone is not sufficient.

（C）BOTH statements TOGETHER are sufficient, but NEITHER statement ALONE is sufficient.

（D）EACH statement ALONE is sufficient.

（E）Statements（1）and（2）TOGETHER are NOT sufficient.

解:

题目问，$n^2 - 1$ 除以 8 的余数是多少。$n^2 - 1$ 可以变为 $(n+1)(n-1)$。

条件 1 说，n 是奇数。如果 n 是奇数，那么 $n+1$ 和 $n-1$ 为两个连续偶数。它们两个的乘积必然为 8 的倍数。故条件 1 充分。

条件 2 说，n 不能被 8 整除。从中我们不能知道 $n-1$ 和 $n+1$ 的乘积与 8 的关系，故条件 2 不充分。

综上，答案为 A。

例题 22

If n is a positive integer and r is the remainder when $(n-1)(n+1)$ is divided by 24, what is the value of r?

(1) n is not divisible by 2.

(2) n is not divisible by 3.

(A) Statement (1) ALONE is sufficient, but statement (2) alone is not sufficient.

(B) Statement (2) ALONE is sufficient, but statement (1) alone is not sufficient.

(C) BOTH statements TOGETHER are sufficient, but NEITHER statement ALONE is sufficient.

(D) EACH statement ALONE is sufficient.

(E) Statements (1) and (2) TOGETHER are NOT sufficient.

解：

题目问的是 $(n-1)(n+1)$ 除以 24 的余数。

条件 1 说，n 不能被 2 整除。这表示，n 是奇数，即 $(n-1)$ 和 $(n+1)$ 是两个连续的偶数。这个条件只能证明 $(n-1)(n+1)$ 是 8 的倍数，但无法证明其是否是 3 的倍数。故条件 1 不充分。

条件 2 说，n 不能被 3 整除。因为 $n-1$，n 和 $n+1$ 为三个连续自然数，所以如果 n 不能被 3 整除，那么 $n-1$ 或 $n+1$ 中的一个必然能被 3 整除。但从这个条件中无法知道 $(n-1)(n+1)$ 是否是 8 的倍数。故条件 2 不充分。

条件 1 + 条件 2，显然，此时 n 既是 8 的倍数，也是 3 的倍数。故条件 1 + 条件 2 充分。

例题 23

If n is an integer, is $\dfrac{n}{15}$ an integer?

(1) $\dfrac{3n}{15}$ is an integer.　　　　(2) $\dfrac{8n}{15}$ is an integer.

(A) Statement (1) ALONE is sufficient, but statement (2) alone is not sufficient.

(B) Statement (2) ALONE is sufficient, but statement (1) alone is not sufficient.

(C) BOTH statements TOGETHER are sufficient, but NEITHER statement ALONE is sufficient.

(D) EACH statement ALONE is sufficient.

(E) Statements (1) and (2) TOGETHER are NOT sufficient.

解：

题目问的是 n 是否能被 15 整除。也就是问 n 通过质因数分解，是否至少能分解出 3 和 5 来。

条件 1 说，$3n$ 能被 15 整除。本条件只能证明 n 进行质因数分解后含有 5，但不能确定是否含有 3。故条件 1 不充分。

条件 2 说，$8n$ 能被 15 整除。本条件能证明 n 进行质因数分解后含必然至少含有一个 3 和一个 5。故条件 2 充分。

综上，答案为 B。

例题 24

If s is an integer, is 24 a divisor of s?

(1) Each of the numbers 3 and 8 is a divisor of s.

(2) Each of the numbers 4 and 6 is a divisor of s.

(A) Statement (1) ALONE is sufficient, but statement (2) alone is not sufficient.

(B) Statement (2) ALONE is sufficient, but statement (1) alone is not sufficient.

(C) BOTH statements TOGETHER are sufficient, but NEITHER statement ALONE is sufficient.

（D）EACH statement ALONE is sufficient.

（E）Statements（1）and（2）TOGETHER are NOT sufficient.

解：

题目问的是，如果 s 是一个整数，24 是否是 s 的因数。而 $24 = 2^3 \times 3$。

条件 1 说，每一个 3~8 之间的数字都是 s 的因数。这表明，s 既是 3 的倍数，也是 8 的倍数。因此，它必然是 24 的倍数。故条件 1 充分。

条件 2 说，每一个 4~6 之间的数字都是 s 的因数。这无法表明 s 进行质因数分解后是否一定含有 3 个 2，故条件 2 不充分。

综上，答案为 A。

例题 25

Rita and Sam play the following game with n sticks on a table. Each must remove 1, 2, 3, 4 or 5 sticks at a time on alternate turns, and no stick that is removed is put back on the table. The one who removes the last stick（or sticks）from the table wins. If Rita goes first, which of the following is a value of n such that Sam can always win no matter how Rita plays？

（A）7　　　　（B）10　　　　（C）11　　　　（D）12　　　　（E）16

解：

这道题目主要需要理解这个游戏的获胜秘诀。因为每一次只能拿走 1~5 根木棍这 5 种情况，所以，在 Rita 进行游戏时，如果桌面上只剩下 5 根或更少的木棍，则 Rita 必胜；如果剩下 6 根，则 Rita 必败（无论 Rita 拿走几根，都会剩下至少 1 根木棍）；如果剩下 7 根，Rita 如果只拿走 1 根，则她必胜。由此可知，只要 Sam 拿走木棍后，剩下 6 根或 6 的倍数根的木棍，则 Sam 必胜（6 的倍数根，无非是多玩几轮而已，最后一轮依然能保证桌面上剩下 6 根木棍）。

综上，只有选项 D 是 6 的倍数，答案为 D。

例题 26

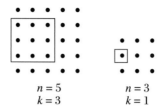

$$n = 5 \qquad n = 3$$
$$k = 3 \qquad k = 1$$

Let n and k be positive integers with $k \leq n$. From an $n \times n$ array of dots, a $k \times k$ array of dots is selected. The figure above shows two examples where the selected $k \times k$ array is enclosed in a square. How many pairs (n, k) are possible so that exactly 48 of the dots in the $n \times n$ array are NOT in the selected $k \times k$ array?

(A) 1   (B) 2   (C) 3   (D) 4   (E) 5

解：

本题最重要的是先看懂题目问的意思。$n \times n$ 是"点"的数量，$k \times k$ 是方框框起来的点的数量。题目问的是，有多少组 n 和 k 的值能使不被框起来的点数恰好是 48。

根据题意，可以列出方程：

$$n^2 - k^2 = 48,$$

即

$$(n - k)(n + k) = 48。$$

由于两个数的乘积是 48，所以我们可以先将 48 进行质因数分解，则有：

$$(n - k)(n + k) = 2^4 \times 3,$$

由此可见，题目即是在问 $2^4 \times 3$ 能构成多少种两个数相乘的模式。因为 $n + k$ 必定大于 $n - k$，所以直接列举时要保证前小后大：

第一组：$1, 2^4 \times 3$；

第二组：$2, 2^3 \times 3$；

第三组：2^2，$2^2 \times 3$；

第四组：3，2^4；

第五组：2×3，2^3。

请注意，这五组还不都满足要求。前者必须能符合 $n-k$，后者必须能符合 $n+k$。通过观察可知，由于 $n-k$ 和 $n+k$ 的和为 $2n$，而 $2n$ 必然为偶数。所以第一组和第四组是不满足条件的，只有 3 组能满足条件。

综上，答案为 C。

例题 27

In an auditorium, 360 chairs are to be set up in a rectangular arrangement with x rows and exactly y chairs each. If the only other restriction is that $10 < x < 25$, how many different rectangular arrangements are possible?

（A）Four　　　（B）Five　　　（C）Six　　　（D）Eight　　　（E）Nine

解：

本题最重要的是把题目的真正意图看懂。一共 360 把椅子，摆成长方形，一共 x 行，y 列。实际上这个问题可以理解为 360 除以 y，商为 x，余数为 0。由此可以联想到，这道题本质上是一个整除问题。问的是 360 能被 10 ~ 25 中的几个数整除。

$$360 = 2 \times 2 \times 2 \times 3 \times 3 \times 5。$$

因此，有如下几组情况是满足条件的：

$(12, 30)$；

$(15, 24)$；

$(18, 20)$；

$(20, 18)$；

$(24, 15)$。

综上，答案为 B。

2.2.3 ▶ 质因数和因数的个数

以 150 为例。请问，150 有多少个质因数和因数呢？

首先，我们需要对 150 做质因数分解，可得：

$$150 = 2 \times 3 \times 5^2。$$

150 中含有的不同的质因数的数量是 3 个，即 2，3 和 5。

其次，在分解质因数后，每项幂指数 +1 再相乘，得到的乘积即为因数个数，也就是能整除 150 的整数个数。150 的因数个数为：$(1+1) \times (1+1) \times (2+1) = 12$ 个。

这个方法的证明可以利用"排列组合"的思想。假设能整除 150 的数是一个容器，

对于 2 来说，这个容器里可以没有，也可以有，共两种可能。

对于 3 来说，这个容器里可以没有，也可以有，共两种可能。

对于 5 来说，这个容器里可以没有，也可以有 1 个，还可以有 2 个，共三种可能。

因此，150 的因数个数为 $2 \times 2 \times 3 = 12$。

如果刚好都没有，则这个因数是 1；如果刚好都有，则这个因数是 150 本身。这两个极值均为 150 的因数，也都是可以整除 150 的数字。

例题 28

How many factors does 735 have?

(A) 11　　　　(B) 12　　　　(C) 13　　　　(D) 14　　　　(E) 15

解：

分解 735 的质因数，则有：

$$735 = 3 \times 5 \times 7^2。$$

735 的因数个数为：$(1+1) \times (1+1) \times (2+1) = 12$ 个。答案为 B。

例题 29

If k is a positive integer, then $20k$ is divisible by how many different positive integers?

(1) k is prime.　　　　　　　(2) $k = 7$

(A) Statement (1) ALONE is sufficient, but statement (2) alone is not sufficient.

(B) Statement (2) ALONE is sufficient, but statement (1) alone is not sufficient.

(C) BOTH statements TOGETHER are sufficient, but NEITHER statement ALONE is sufficient.

(D) EACH statement ALONE is sufficient.

(E) Statements (1) and (2) TOGETHER are NOT sufficient.

解：

题目问的是 $20k$ 能被多少个整数整除，这个问题本质上就是因数个数问题。$20k = 2^2 \times 5 \times k$。我们要考虑 k 的质因数情况。

条件 1 说，k 是质数。请注意，质数也有偶数的情况，比如 k 是 2，那么 $20k$ 的因数个数为 $4 \times 2 = 8$ 个；而再比如 k 是 5，则 $20k$ 的因数个数为 $3 \times 3 = 9$ 个。因此，条件 1 不充分。

条件 2 说，$k = 7$。显然，如果给出 k 的具体数字，那么一定可以确定 $20k$ 的质因数分解情况，进而确定因数个数，故条件 2 充分。

综上，答案为 B。

例题 30

If the prime numbers p and t are the only prime factors of the integer m, is m a multiple of $p^2 t$?

(1) m has more than 9 positive factors.

(2) m is a multiple of p^3.

(A) Statement (1) ALONE is sufficient, but statement (2) alone is not sufficient.

(B) Statement (2) ALONE is sufficient, but statement (1) alone is not sufficient.

(C) BOTH statements TOGETHER are sufficient, but NEITHER statement ALONE is sufficient.

(D) EACH statement ALONE is sufficient.

(E) Statements (1) and (2) TOGETHER are NOT sufficient.

解：

m 有两个质因数，一个是 p 一个是 t。题目问的是，m 是否含有两个 p 和一个 t。其实我们只需要知道 m 进行质因数分解后是否含有两个 p 即可（t 在题干中已默认有一个了）。

条件 1 说，m 有 9 个正因数。无论 m 有几个因数，这些因数中可能只有一个 p，其他都是 t。因此，条件 1 不充分。

条件 2 说，m 是 p^3 的倍数。显然，这个条件告诉我们 m 进行质因数分解后至少含有 3 个 p，故条件 2 充分。

综上，答案为 B。

2.2.4 ▶ 最大公约数和最小公倍数

求几个数的最大公约数：把每个数分别分解质因数，再把各数中的全部公有质因数提取出来连乘，所得的积就是这几个数的最大公约数（如果各数中的质因数有相同的情况，则比较各数中哪个数有该质因数的个数较少，乘较少的次数）。

求几个数的最小公倍数：先把这几个数的质因数写出来，最小公倍数等于它们所有的质因数的乘积（如果各数中的质因数有相同的情况，则比较各数中哪个数有该质因数的个数较多，乘较多的次数）。

例如：求 30 和 45 的最大公约数和最小公倍数。

先将 30 和 45 进行质因数分解：

$$30 = 2 \times 3 \times 5,$$
$$45 = 3 \times 3 \times 5 = 3^2 \times 5。$$

根据定义，则有：

30 和 45 的最大公约数：$3 \times 5 = 15$，

30 和 45 的最小公倍数：$2 \times 3 \times 3 \times 5 = 90$。

例题 31

If x and y are positive integers, what is the value of xy?

(1) The greatest common factor of x and y is 10.

(2) The least common multiple of x and y is 180.

(A) Statement (1) ALONE is sufficient, but statement (2) alone is not sufficient.

(B) Statement (2) ALONE is sufficient, but statement (1) alone is not sufficient.

(C) BOTH statements TOGETHER are sufficient, but NEITHER statement ALONE is sufficient.

(D) EACH statement ALONE is sufficient.

(E) Statements (1) and (2) TOGETHER are NOT sufficient.

解：

本题已知最大公约数和最小公倍数，要求两个数的乘积，已知最大公约数就是两个数指数较小的公因数的乘积，最小公倍数就是公因数中指数较大的以及其他所有非公因数的乘积。

仅知道公倍数和公约数明显无法得到充分结论，所以我们直接来看二者相结合的情况。

最大公约数是 10，所以我们知道 x 和 y 仅有两个公因数：2 和 5，

最小公倍数是 180，$180 = 2 \times 2 \times 3 \times 3 \times 5$。

所以我们得到结论，3×3 是所有的非公因数，仅出现在了 x 或 y 某一个数字的因数中，5 是公因数，在两个数字的因数中都只出现了 1 次，因数 2 由于在最小公倍数中出现了 2 次，但在最大公约数中只出现了一次，所以因数 2 分别在 x 和 y 这两个数字的因数中出现了 1 次和 2 次。

结合上面的结论，最终 $xy = 2 \times 2 \times 2 \times 3 \times 3 \times 5 \times 5$。

综上，答案为 C。

例题 32

If n and t are positive integers, what is the greatest prime factor of the product nt?

(1) The greatest common factor of n and t is 5.

(2) The least common multiple of n and t is 105.

(A) Statement (1) ALONE is sufficient, but statement (2) alone is not sufficient.

(B) Statement (2) ALONE is sufficient, but statement (1) alone is not sufficient.

(C) BOTH statements TOGETHER are sufficient, but NEITHER statement ALONE is sufficient.

(D) EACH statement ALONE is sufficient.

(E) Statements (1) and (2) TOGETHER are NOT sufficient.

解:

本题问的是 $n \times t$ 的最大质因数，也就是问 $n \times t$ 的质因数分解。

条件1说，n 和 t 的最大公约数是 5，从中我们无法知道 n 或者 t 是否有更大的质因数，故条件1不充分。

条件2说，n 和 t 的最小公倍数是 105。根据最小公倍数的定义，n 和 t 的所有质因数都应该出现在两者的最小公倍数中，自然能从中得知 $n \times t$ 的最大质因数。故条件2充分。

综上，答案为 B。

例题 33

If x is a positive integer, what is the least common multiple of x, 6, and 9?

(1) The least common multiple of x and 6 is 30.

(2) The least common multiple of x and 9 is 45.

(A) Statement (1) ALONE is sufficient, but statement (2) alone is not sufficient.

(B) Statement (2) ALONE is sufficient, but statement (1) alone is not sufficient.

(C) BOTH statements TOGETHER are sufficient, but NEITHER statement ALONE is sufficient.

（D）EACH statement ALONE is sufficient.

（E）Statements（1）and（2）TOGETHER are NOT sufficient.

解：

题目问的是三个数的最小公倍数。因此我们需要知道 x 的质因数分解模式。

条件 1 说，x 和 6 的最小公倍数是 $2 \times 3 \times 5$。由此可知，x 分解质因数后至少含有一个 5，至多则含有一个 2，一个 3 和一个 5。因此，无论 x 是至多还是至少或者是处于中间状态，我们都可以确定 x，6，9 的最小公倍数必然为 $2 \times 3^2 \times 5$。故条件 1 充分。

条件 2 说，x 和 9 的最小公倍数是 $3^2 \times 5$。由此可知，x 分解质因数后至少含有一个 5，至多含有两个 3 和一个 5。因此，无论 x 是至多还是至少或者是处于中间状态，x，6，9 的最小公倍数都必然为 $2 \times 3^2 \times 5$。故条件 2 充分。

综上，答案为 D。

例题 34

What is the greatest number of identical bouquets that can be made out of 21 white and 91 red tulips if no flowers are to be left out?（Two bouquets are identical whenever the number of red tulips in the two bouquets is equal and the number of white tulips in the two bouquets is equal.）

（A）3 　　　　（B）4 　　　　（C）5 　　　　（D）6 　　　　（E）7

解：

这道题的主要难点在于理解它的实质上考的是最大公约数的问题。

题目问的是，如果没有花剩下的话，那么最多能有多少束相同的花。因为相同的花的定义是，两种花束中白色花的数量相等，红色花的数量也相等。所以，我们直接求 21 和 91 的最大公约数。

$$21 = 3 \times 7, \ 91 = 7 \times 13；$$

两个数的最大公约数为 7。

这个意思是，我们可以把白花分成 7 组，每组 3 朵；把红花也分为 7 组，每组 13 朵。由此可以构成 7 束花，每束花均有 3 朵白花和 13 朵红花。这 7 束花完全相同。

综上，答案为 E。

2.3 ▸ 余数

如果 x 和 y 是两个正整数，则会有两个非常独特的整数，分别为 q 和 r，一个叫商（quotient），一个叫余数（remainder），可以写成如下的表达式：

$$y = xq + r \quad (0 \leqslant r < x)$$

例如，$28 = 8 \times 3 + 4$。这个式子表示，28 除以 8，商是 3，余数是 4。如果被除数（divisor）比余数小，例如 5 除以 7，则商为 0，余数为 5。

要想充分理解余数问题，需要先理解除法的本质。除法的本质是"排队"。例如，10 除以 4 的实际意义是，总共有 10 个球进行排队，要求每一行放置 4 个球，则我们必然会排成下图：

此时，我们能排出两个完整的行。有几个完整的行，商即为几。由于还剩下两个球无法构成完整行，所以那两个球即为余数。因此，10 除以 4，商为 2，余数为 2。

我们可以利用除法的本质来回答在整除和余数考题中常见的两个现象。

1 如果 n 能被某数整除，那么 $n+1$ 就必然不能被这个数整除。

试想，如果 n 能被某数整除，则证明把 n 个球按照该数一行的方式排队，不会有余下的球。例如，如果 n 能被 3 整除，则将 n 按照 3 个一行进行排队，必然可以排满，没有剩余。

此时，如果再加上一个球，则有：

多加的那个球必然是余数，表明 $n+1$ 不能再被 3 整除。同理，$n+2$ 也不能被 3 整除。

例题 1

If x is an integer and $y = 3x + 2$, which of the following CANNOT be a divisor of y?

(A) 4   (B) 5   (C) 6   (D) 7   (E) 8

解：

根据题干，$y = 3x + 2$。

因为 $3x$ 必然是 3 的倍数，一个数是 3 的倍数加 2，表明它一定不能是 3 的倍数。因此，y 一定不能被 3 的倍数整除，故答案为 C。

例题 2

For every positive even integer n, the function $h(n)$ is defined to be the product of all the even integers from 2 to n, inclusive. If p is the smallest prime factor of $h(100) + 1$, then p is

(A) between 2 and 10.   (B) between 10 and 20.

(C) between 20 and 30.   (D) between 30 and 40.

(E) greater than 40.

解：

这道题看起来很难，其实理解了 n 和 $n+1$ 的原理后会非常简单。显然，基于 $h(n)$ 函数的定义，$h(100)$ 包含了所有 100 以内偶数的乘积。也就是说，所有 50 以内的质数，一定都是 $h(100)$ 的因数。这是因为，50 以内的任何数乘 2 后均是一个小于 100 的偶数，它必然被包含在 $h(100)$ 内。

因为 50 以内的任何质数都是 $h(100)$ 的因数，所以它们一定不是 $h(100) + 1$ 的因数。也就是说，最小的 $h(100) + 1$ 的因数一定会超过 50，那就更会超过 40 了。

综上，答案为 E。

2 3 个连续正整数的乘积必然是 6 的倍数；2 个连续偶数的乘积必然是 8 的倍数。

先来证明 3 个连续的正整数乘积必然是 6 的倍数。

首先，任意 3 个连续的正整数，其中必然有至少一个偶数，这是不证自明的，因为每隔一个数，就会出现一个偶数。一个偶数中至少含有一个 2。

其次，任意 3 个连续的正整数，无非是下列三种情况，如下图：

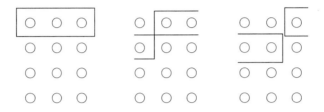

框内的 3 个球，表示任意 3 个连续自然数。我们可以看到，只要是 3 个连续的自然数，必然包含最右侧一列数中的一个，即

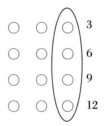

因此，可以证明，任意 3 个连续的自然数，其中一定有一个能被 3 整除。

由此可知，3 个连续正整数的乘积必然是 6 的倍数。

证明"2 个连续偶数的乘积必然是 8 的倍数"的方法与上述证明过程相似。

因为两个连续的偶数必然出现在连续的 4 个整数中。由刚才的证明可知，任意连续的 4 个整数，一定有一个能被 4 整除。因此，两个连续的偶数，必然至少一个能被 2 整除，一个能被 4 整除。也就是说，2 个连续偶数的乘积必然是 8 的倍数。

⟐ 高斯同余定理

余数问题的本质是高斯同余定理问题。如果一个余数的问题不能用高斯同余定理来解，那么这道余数题就永远没有解。

高斯同余定理的形式非常简单，分别是加、减、乘三项定理：

若 a 和 a' 除以 m 余数相同，且 b 和 b' 除以 m 余数也相同，则 $a+b$ 和 $a'+b'$ 除以 m，余数依然相同。例如：

因为 1 和 11 除以 10 都余 1；3 和 13 除以 10 都余 3，所以 $1+3$ 除以 10 的余数等于 $11+13$ 除以 10 的余数。

为了方便书写，我们可以把这条定理写为：

如果 $r(a)=r(a')(\bmod m)$，$r(b)=r(b')(\bmod m)$，那么 $r(a+b)=r(a'+b')(\bmod m)$。

其中，$r(a)$ 表示 a 除以 m 后的余数；$\bmod m$ 是除以 m 的意思。

高斯同余定理的加减乘定理为：

若 $r(a)=r(a')(\bmod m)$，$r(b)=r(b')(\bmod m)$，

（1）$r(a)\pm r(b)=r(a')\pm r(b')(\bmod m)$；

（2）$r(a)\times r(b)\equiv r(a')\times r(b')(\bmod m)$。

而在求解余数问题时，我们更多用到的是高斯同余定理的推论。请大家记住下面两个公式。

（1）$r(a)\pm r(b)=r(a\pm b)$

（2）$r(a)\times r(b)=r(a\times b)$

也就是说，两个数相加（或相减、相乘）后除以某个数的余数，等于这两个数分别除以这个数后的余数再相加（或相减、相乘）。

这个定理在很多我们没有意识到的地方发挥着重要的作用。例如，请判断：

2134 是否能被 3 整除？

很多考生会马上想到，只要检查 2134 中的各个数位相加之后的和是否能被 3 整除即可。

那么，再请问，为什么？

相信这个问题，大部分考生就答不上来了。其实，它是高斯同余定理的一个简单应用。

请大家跟着我在纸上用手写一遍这个证明过程，加深对高斯同余定理的印象。

我们可以把2134先写成：

$$2000 + 100 + 30 + 4。$$

根据高斯同余的加法定理，求2134除以3的余数，就是求这四个数分别除以3的余数之后再相加，即：

$$r(2000 + 100 + 30 + 4) = r(2000) + r(100) + r(30) + r(4) = r(2 \times 1000) + r(1 \times 100) + r(3 \times 10) + r(4 \times 1)。$$

根据高斯同余乘法定理，则有：

$$r(2 \times 1000) + r(1 \times 100) + r(3 \times 10) + r(4 \times 1)$$
$$= r(2) \times r(1000) + r(1) \times r(100) + r(3) \times r(10) + r(4) \times r(1)$$

我们知道，无论10的多少次方除以3，它余数都为1，因此：

$$r(2) \times r(1000) + r(1) \times r(100) + r(3) \times r(10) + r(4) \times r(1) = r(2) + r(1) + r(3) + r(4)。$$

再用一次高斯同余的加法定理，则有：

$$r(2) + r(1) + r(3) + r(4) = r(2 + 1 + 3 + 4)。$$

由此我们利用高斯同余定理证明了，$r(2000 + 100 + 30 + 4) = r(2 + 1 + 3 + 4)$，即2134除以3的余数与 $2 + 1 + 3 + 4$ 除以3的余数相同。

因此，如果 $r(2 + 1 + 3 + 4) = 0$，即如果2134中的各个数位相加之后的和能被3整除，那么2134就能被3整除。

熟练应用高斯同余定理会帮助我们解决所有GMAT考试中的余数问题，让我们看一些例题。

例题 3

If n is a positive integer and r is the remainder when $4 + 7n$ is divided by 3, what is the value of r？

（1）$n + 1$ is divisible by 3.

（2）$n > 20$

（A）Statement（1）ALONE is sufficient, but statement（2）alone is not sufficient.

（B）Statement（2）ALONE is sufficient, but statement（1）alone is not sufficient.

（C）BOTH statements TOGETHER are sufficient, but NEITHER statement ALONE is sufficient.

（D）EACH statement ALONE is sufficient.

（E）Statements（1）and（2）TOGETHER are NOT sufficient.

解：

题目问的是，$4+7n$ 除以 3 的余数为几。根据高斯同余加法和乘法定理可知：

$$r(4+7n) = r(4) + r(7n) = r(4) + r(7) \times r(n) = 1 + 1 \times r(n) = 1 + r(n)。$$

因此，想知道 $4+7n$ 除以 3 的余数，只需要知道 n 除以 3 的余数即可。

条件 1 说，$n+1$ 是可以被 3 整除的。由此可知，n 除以 3 必然余 2。故条件 1 充分。

条件 2 说，n 大于 20，显然无法得知 n 除以 3 的余数，故条件 2 不充分。

综上，答案为 A。

例题 4

What is the remainder when the positive integer n is divided by 5?

（1）When n is divided by 3, the quotient is 4 and the remainder is 1.

（2）When n is divided by 4, the remainder is 1.

（A）Statement（1）ALONE is sufficient, but statement（2）alone is not sufficient.

（B）Statement（2）ALONE is sufficient, but statement（1）alone is not sufficient.

（C）BOTH statements TOGETHER are sufficient, but NEITHER statement ALONE is sufficient.

（D）EACH statement ALONE is sufficient.

（E）Statements（1）and（2）TOGETHER are NOT sufficient.

解：

题目问的是，n 被 5 除后余几。

条件 1 说，n 除以 3，商为 4，余数为 1。通过条件 1，我们显然可以直接知道 n 等于几，进而必然可以判断 n 除以 5 的余数，故条件 1 充分。

条件 2 说，n 除以 4，余数为 1。而最小的能除以 4 余 1 的数是 1（因为 1 小于 4，所以余数是 1）。根据高斯同余加法定理可知：

$$r(1) = r(1) + 0 = r(1) + r(4k) = r(1 + 4k) \ (k = 0, 1, 2 \cdots)。$$

这是因为，$4k$ 必然是 4 的倍数，所以 $r(4k)$ 等于 0。

因此，所有的 $1 + 4k$ 的数，均可以满足除以 4 余 1。

此时，n 可以等于 1，也可以等于 5，不能确定 n 除以 5 的余数。故条件 2 不充分。

综上，答案为 A。

例题 5

If x is an integer greater than 0, what is the remainder when x is divided by 4?

(1) The remainder is 3 when $x + 1$ is divided by 4.

(2) The remainder is 0 when $2x$ is divided by 4.

(A) Statement (1) ALONE is sufficient, but statement (2) alone is not sufficient.

(B) Statement (2) ALONE is sufficient, but statement (1) alone is not sufficient.

(C) BOTH statements TOGETHER are sufficient, but NEITHER statement ALONE is sufficient.

(D) EACH statement ALONE is sufficient.

(E) Statements (1) and (2) TOGETHER are NOT sufficient.

解：

题目问的是，x 除以 4 余几。

条件 1 说，当 $x + 1$ 除以 4 的时候，余数为 3。根据高斯同余定理中的加法定理，则有：

$$r(x + 1) = r(x) + r(1) = r(x) + 1 = 3。$$

则 $r(x) = 2$；故条件 1 充分。

条件 2 说，$2x$ 除以 4 余数为 0，我们只能确定 x 是 2 的倍数，不能确定 x 是否是 4 的倍数，故条件 2 不充分。

综上，答案为 A。

例题 6

When 24 is divided by the positive integer n, the remainder is 4. Which of the following statements about n must be true?

(I) n is even.

(II) n is a multiple of 5.

(III) n is a factor of 20.

(A) III only

(B) I and II only

(C) I and III only

(D) II and III only

(E) I , II , and III

解：

我们可以把 n 写为：

$24 = q \times n + 4$（$q = 1,\ 2 \cdots$）；即 $q \times n = 20 = 2^2 \times 5$。

因此，从表面上看起来，n 可以等于 1，2，4，5，10，20。

但是，请注意，如果 $n = 1, 2$ 或 4，此时 24 是可以被 n 整除的，不符合题设。因此，n 只能为 5，10，20。

显然，答案为 D。

What is the remainder when the positive integer n is divided by 3?

(1) The remainder when n is divided by 2 is 1.

(2) The remainder when $n+1$ is divided by 3 is 2.

(A) Statement (1) ALONE is sufficient, but statement (2) alone is not sufficient.

(B) Statement (2) ALONE is sufficient, but statement (1) alone is not sufficient.

(C) BOTH statements TOGETHER are sufficient, but NEITHER statement ALONE is sufficient.

(D) EACH statement ALONE is sufficient.

(E) Statements (1) and (2) TOGETHER are NOT sufficient.

解:

题目问的是 n 除以 3 的余数为几。

条件 1 说，n 除以 2 余 1。显然，n 除以 2 的余数与 n 除以 3 的余数无关。故条件 1 不充分。

条件 2 说，$n+1$ 除以 3 余 2。根据高斯同余的加法定理可知:

$$r(n+1) = r(n) + r(1) = 2,$$

由此可知,

$$r(n) = 1。$$

故条件 2 充分。

综上，答案为 B。

If x and y are positive integers, what is the remainder when $10^x + y$ is divided by 3 ?

(1) $x = 5$

(2) $y = 2$

(A) Statement (1) ALONE is sufficient, but statement (2) alone is not sufficient.

（B）Statement（2）ALONE is sufficient, but statement（1）alone is not sufficient.

（C）BOTH statements TOGETHER are sufficient, but NEITHER statement ALONE is sufficient.

（D）EACH statement ALONE is sufficient.

（E）Statements（1）and（2）TOGETHER are NOT sufficient.

解：

题目问的是，$10^x + y$ 除以 3 的余数为几。根据高斯同余定理的加法定理可知，

$$r(10^x + y) = r(10^x) + r(y)。$$

无论是 10 的多少次方，它除以 3 的余数一定为 1。

关于这一点，也可以用高斯同余定理的乘法定理来证明。我们知道，10 除以 3 余数为 1，即 10 除以 3 的余数和 1 除以 3 的余数相同，可以写为：

$$r(10) = r(1)，$$

让等式两边同时乘以自己，可得：

$$r(10) \times r(10) = r(1) \times r(1)，$$

即

$$r(100) = r(1)，$$

以此类推，无论 10 的多少次方，永远和 1 同余。

因此，

$$r(10^x) + r(y) = 1 + r(y)，$$

也就是说，我们只需要知道 y 是几，即可以知道问题的答案。

综上，答案为 B。

例题 9

If t is a positive integer and r is the remainder when $t^2 + 5t + 6$ is divided by 7, what is the value of r?

(1) When t is divided by 7, the remainder is 6.

(2) When t^2 is divided by 7, the remainder is 1.

(A) Statement (1) ALONE is sufficient, but statement (2) alone is not sufficient.

(B) Statement (2) ALONE is sufficient, but statement (1) alone is not sufficient.

(C) BOTH statements TOGETHER are sufficient, but NEITHER statement ALONE is sufficient.

(D) EACH statement ALONE is sufficient.

(E) Statements (1) and (2) TOGETHER are NOT sufficient.

解:

题目问的是 $t^2 + 5t + 6$ 除以 7 的余数为几。

条件 1 说，t 除以 7 的时候，余数为 6。显然，根据高斯同余定理的乘法定理，知道 $r(t)$ 等于知道 $r(t^2)$。因此，$t^2 + 5t + 6$ 除以 7 的余数可以计算出来。故条件 1 充分。

条件 2 说，t^2 除以 7，余数为 1。

这个条件有些难度，乍看之下也是充分的，但我们需要仔细求证。根据高斯同余乘法定理，则有:

$$r(t^2) = r(t) \times r(t) = r(1) = 1,$$

如果 t 除以 7 的余数为 1，那么 $r(t) \times r(t) = 1$;

如果 t 除以 7 的余数为 2，那么 $r(t) \times r(t) = r(4) = 4$;

如果 t 除以 7 的余数为 3，那么 $r(t) \times r(t) = r(9) = 2$，由于 9 比 7 大，可以再排一队（除法的本质是排队，上文讲过），导致余数为 2;

如果 t 除以 7 的余数为 4，那么 $r(t) \times r(t) = r(16) = 2$;

如果 t 除以 7 的余数为 5，那么 $r(t) \times r(t) = r(25) = 4$；

如果 t 除以 7 的余数为 6，那么 $r(t) \times r(t) = r(36) = 1$。

显然，t 除以 7 余 1 和 t 除以 7 余 6 时，t^2 除以 7 的余数均为 1。因此，无法确定 t 除以 7 的余数。故条件 2 不充分。

综上，答案为 A。

例题 10

What is the remainder when the positive integer n is divided by the positive integer k, where $k > 1$?

(1) $n = (k+1)^3$ (2) $k = 5$

(A) Statement (1) ALONE is sufficient, but statement (2) alone is not sufficient.

(B) Statement (2) ALONE is sufficient, but statement (1) alone is not sufficient.

(C) BOTH statements TOGETHER are sufficient, but NEITHER statement ALONE is sufficient.

(D) EACH statement ALONE is sufficient.

(E) Statements (1) and (2) TOGETHER are NOT sufficient.

解：

题目问的是，n 除以 k 的余数为几。

由条件 1 可知，在 n 除以 k 时，$r(n) = r((k+1)^3)$。根据高斯同余定理的乘法定理可知，

$$r((k+1)^3) = r(k+1) \times r(k+1) \times r(k+1)。$$

根据高斯同余乘法定理可知，

$$r(k+1) = r(k) + r(1) = 0 + 1 = 1。$$

因此，$r(n) = 1$。故条件 1 充分。

条件 2 只给出了 k 的值，无法得知 n 除以 k 的余数为几，故条件 2 不充分。

综上，答案为 A。

例题 11

If x, y, and z are positive integers, what is the remainder when $100x + 10y + z$ is divided by 7 ?

(1) $y = 6$

(2) $z = 3$

(A) Statement (1) ALONE is sufficient, but statement (2) alone is not sufficient.

(B) Statement (2) ALONE is sufficient, but statement (1) alone is not sufficient.

(C) BOTH statements TOGETHER are sufficient, but NEITHER statement ALONE is sufficient.

(D) EACH statement ALONE is sufficient.

(E) Statements (1) and (2) TOGETHER are NOT sufficient.

解：

题目问的是，$100x + 10y + z$ 除以 7 的余数为几。根据高斯同余定理可知：

$$
\begin{aligned}
& r(100x + 10y + z) \\
&= r(100x) + r(10y) + r(z) \\
&= r(100) \times r(x) + r(10) \times r(y) + r(z) \\
&= 2 \times r(x) + 3 \times r(y) + r(z)
\end{aligned}
$$

依然需要知道 x，y，z 分别除以 7 的余数为几。

显然，条件 1 + 条件 2 都不够充分。综上，答案为 E。

例题 12

If p and n are positive integers and $p > n$, what is the remainder when $p^2 - n^2$ is divided by 15 ?

(1) The remainder when $p + n$ is divided by 5 is 1.

(2) The remainder when $p - n$ is divided by 3 is 1.

(A) Statement (1) ALONE is sufficient, but statement (2) alone is not sufficient.

(B) Statement (2) ALONE is sufficient, but statement (1) alone is not sufficient.

（C）BOTH statements TOGETHER are sufficient，but NEITHER statement ALONE is sufficient.

（D）EACH statement ALONE is sufficient.

（E）Statements（1）and（2）TOGETHER are NOT sufficient.

解：

题目问的是 $p^2 - n^2$ 除以 15 后的余数为几。显然，想知道这个，就是要知道 $(p+n) \times (p-n)$ 除以 15 后的余数为几。根据高斯同余乘法定理，我们只需要知道 $p+n$ 和 $p-n$ 分别除以 15 后的余数为几即可。

条件 1 告诉我们的是 $p+n$ 除以 5 后的余数，但不能知道 $p+n$ 除以 15 后的余数，故条件 1 不充分。

条件 2 告诉我们的是 $p-n$ 除以 3 后的余数，但不能知道 $p-n$ 除以 15 后的余数，故条件 2 不充分。

两者相加，亦不能知道 $(p+n) \times (p-n)$ 除以 15 后的余数，因此条件 1 + 条件 2 也不充分。

综上，答案为 E。

例题 13

What is the value of the positive integer m?

（1）When m is divided by 6, the remainder is 3.

（2）When 15 is divided by m, the remainder is 6.

（A）Statement（1）ALONE is sufficient，but statement（2）alone is not sufficient.

（B）Statement（2）ALONE is sufficient，but statement（1）alone is not sufficient.

（C）BOTH statements TOGETHER are sufficient，but NEITHER statement ALONE is sufficient.

（D）EACH statement ALONE is sufficient.

（E）Statements（1）and（2）TOGETHER are NOT sufficient.

解：

条件 1 说，m 除以 6 余数为 3。除以 6 后余数为 3 的数很多，3，9，15 等均是，故条件 1 不充分。

条件 2 说，15 除以 m 余数为 6。余数肯定要小于除数，所以 m 必然大于 6。大于 6 的能使 15 被除后余 6 的数，只有 9。故条件 2 充分。

综上，答案为 B。

有五种经典题型考查高斯同余定理，它们分别为：

（1）余数公式
（2）特殊数字整除
（3）大数求余
（4）中国剩余定理
（5）万年历问题

 余数公式

余数公式是 $y = q \times x + r$。有些考题会直接考查考生对于这个公式的理解和把握。

例题 14

When the integer n is divided by 17, the quotient is x and the remainder is 5. When n is divided by 23, the quotient is y and the remainder is 14. Which of the following is true?

(A) $23x + 17y = 19$ (B) $17x - 23y = 9$ (C) $17x + 23y = 19$

(D) $14x + 5y = 6$ (E) $5x - 14y = -6$

解：
依题意，得出：$n = 17x + 5$；$n = 23y + 14$。

联立两者可得：

$$17x + 5 = 23y + 14,$$

即

$$17x - 23y = 9。$$

答案为 B。

例题 15

When the positive integer x is divided by the positive integer y, the remainder is 9. If $\dfrac{x}{y} = 96.12$, what is the value of y?

（A）96 （B）75 （C）48 （D）25 （E）12

解：

依题意，$x = q \times y + 9$；$x = 96.12y$。

仅凭借这两个方程是无法求 x, y 和 q 这三个未知数的。

根据余数定义可知，如果 x 除以 y 等于 96.12，这表示，x 个球，按照 y 个球一行进行排队，是可以排出 96 行还多出一些球的。96 个完整行就是商，0.12 对应的就是余数。联立方程：

$$x = 96y + 9,$$
$$x = 96.12y,$$

则，$y = 75$。答案为 B。

例题 16

If r is the remainder when the positive integer n is divided by 7, what is the value of r?

（1）When n is divided by 21, the remainder is an odd number.

（2）When n is divided by 28, the remainder is 3.

（A）Statement（1）ALONE is sufficient, but statement（2）alone is not sufficient.

（B）Statement（2）ALONE is sufficient, but statement（1）alone is not sufficient.

（C）BOTH statements TOGETHER are sufficient, but NEITHER statement ALONE is sufficient.

（D）EACH statement ALONE is sufficient.

（E）Statements（1）and（2）TOGETHER are NOT sufficient.

解：

题目问的是，n 除以 7，余数为几。两个条件给出的都是 n 除以其他数的余数。原则上，我们无法判断 n 除以不同数后的余数关系。但有一种情况例外，当除数之间是倍数关系时，利用余数公式和高斯同余定理，可以得到余数的关系。

条件 1 说，n 除以 21，余数是奇数。显然，只告诉我们余数是奇数，无法判断余数具体是几，故条件 1 不充分。

条件 2 说，n 除以 28，余数是 3。可以看到，28 是 7 的倍数。由此，我们可以将 n 写为：

$$n = 28k + 3。$$

根据高斯同余加法定理，可得出：

$$r(n) = r(28k) + r(3)。$$

因为 28 是 7 的倍数，所以 $r(28k) = 0$。

由此可知，

$$r(n) = 0 + r(3) = 3。$$

故条件 2 充分。

综上，答案为 B。

例题 17

If x and y are integers, is $xy + 1$ divisible by 3?

(1) When x is divided by 3, the remainder is 1.

(2) When y is divided by 9, the remainder is 8.

(A) Statement (1) ALONE is sufficient, but statement (2) alone is not sufficient.

(B) Statement (2) ALONE is sufficient, but statement (1) alone is not sufficient.

(C) BOTH statements TOGETHER are sufficient, but NEITHER statement ALONE is sufficient.

(D) EACH statement ALONE is sufficient.

(E) Statements (1) and (2) TOGETHER are NOT sufficient.

解:

题目问的是，$xy + 1$ 是否能被 3 整除。根据高斯同余加法定理可知，

$$r(xy + 1) = r(xy) + r(1) = r(xy) + 1。$$

也就是说，题目的最简需求是计算 $r(xy)$ 是否等于 2。

(这是因为，如果两个数的余数相加是 3 的倍数，则表示可以继续排完整的行。即一个数除以 3 余 3，等于这个数除以 3 余 0。)

条件 1 说，x 除以 3 余 1。由于不知道 y 的情况，故条件 1 不充分。

条件 2 说，y 除以 9 余 8。由于不知道 x 的情况，故条件 2 不充分。

条件 1 + 条件 2，我们可以把 y 表达成 $9k + 8$。根据高斯同余加法定理，$r(9k + 8) = r(9k) + r(8)$。显然 $9k$ 必为 3 的倍数，因此，$r(9k + 8) = 0 + r(8) = 2$。

已知 x 和 y 分别除以 3 的余数，必然能知道 xy 除以 3 的余数，进而知道 $r(xy)$ 是否等于 2。

因此，答案为 C。

例题 18

What is the remainder when the positive integer x is divided by 3?

(1) When x is divided by 6, the remainder is 2.

(2) When x is divided by 15, the remainder is 2.

(A) Statement (1) ALONE is sufficient, but statement (2) alone is not sufficient.

(B) Statement (2) ALONE is sufficient, but statement (1) alone is not sufficient.

(C) BOTH statements TOGETHER are sufficient, but NEITHER statement ALONE is sufficient.

(D) EACH statement ALONE is sufficient.

(E) Statements (1) and (2) TOGETHER are NOT sufficient.

解：

题目问 x 除以 3 余数为几。

条件 1 说，当 x 被 6 除的时候，余数为 2。根据余数公式，可知：$x = 6k + 2$。因此，当 x 除以 3 时，根据高斯同余加法定理，则有：

$$r(x) = r(6k + 2) = r(6k) + r(2),$$

因为 $6k$ 必然是 3 的倍数，所以，

$$r(6k) + r(2) = 0 + 2 = 2,$$

故条件 1 充分。

条件 2 与条件 1 同理。

综上，答案为 D。

---

2 特殊数字整除

根据高斯同余定理，我们很容易得出一些数字的整除规律。在之前的章节中已经证明过除以 3 的整除规律了，其他整除规律的证明与之大同小异。为了能在考试中快速解题，大

家可以熟记下列几个常见的特殊数字整除规律。

(1) 2 的倍数特征：个位数字是偶数。

(2) 5 的倍数特征：个位数字是 0 或者 5。

(3) 3 或 9 的倍数特征：各个数位之和能被 3 或 9 整除。

(4) 4 的倍数特征：末两位数能被 4 整除。

(5) 8 的倍数特征：末三位数能被 8 整除。

(6) 25 的倍数特征：末两位数能被 25 整除。

(7) 125 的倍数特征：末三位数能被 125 整除。

例题 19

If a six-digit number $6ab11c$ is divisible by 8, $c =$

(A) 1   (B) 2   (C) 3   (D) 4   (E) 5

解：

8 的倍数特征为：末三位数能被 8 整除，$112 = 8 \times 14$。因此，$c = 2$，答案为 B。

例题 20

What is the remainder when the two-digit, positive integer x is divided by 3?

(1) The sum of the digits of x is 5.

(2) The remainder when x is divided by 9 is 5.

(A) Statement (1) ALONE is sufficient, but statement (2) alone is not sufficient.

(B) Statement (2) ALONE is sufficient, but statement (1) alone is not sufficient.

(C) BOTH statements TOGETHER are sufficient, but NEITHER statement ALONE is sufficient.

(D) EACH statement ALONE is sufficient.

(E) Statements (1) and (2) TOGETHER are NOT sufficient.

解:

题目问的是,两位数 x 除以 3 余数为几。

条件 1 说,x 的两个数位相加是 5。我们在高斯同余定理部分举例时讲过一个数除以 3 的特征,很容易证明,两位数除以 3 的余数和这个两位数的两个数位相加后除以 3 的余数相等。故条件 1 充分。

条件 2 说,x 除以 9 余数为 5。x 可以写为 $9k+5$,除以 3 后的余数可以写为:

$$r(9k+5) = r(9k) + r(5) = 0 + r(5) = 2。$$

故条件 2 充分。

综上,答案为 D。

---

③ 大数求余

这里谈到的大数,可不是几百几千,而是那些连普通计算器都算不出来的大数,抑或是本身就无法确定具体数值的数(比如带有未知数 x)。比如,如果问 7 除以 10 余数是多少?相信大家一定能很快得出答案。但如果问,7^{548} 除以 10 余数是多少?想必大家就需要思考一阵子了。这里的 7^{548} 就是我们说的"大数"。

在"大数求余"类的考题中,高斯同余的乘法定理会被反复使用。方法分为以下两步:

例如:7^{548} 除以 10 余数为几?

第一步:

先通过"试数"的方式,寻找 7 的多少次方除以 10 余 1(原则上,可以通过费马定理求解,但 GMAT 不会涉及很难的数字,所以直接试数即可)。

7^1 除以 10 余 7;7^2 除以 10 余 9;7^3 除以 10 余 3;7^4 除以 10 余 1。

找到余 1 的情况,即 7 的 4 次方。

第二步：

利用高斯同余乘法定理解题。

既然 7 的 4 次方除以 10 余 1，由于 1 除以 10 也余 1，所以在同除以 10 的情况下，7^4 和 1 是同余的。可以表达为：

$$r(7^4) = r(1)。$$

等式左边乘以 r (7^4)，右边乘以 r (1)，两者一定依然相等，即

$$r(7^4) \times r(7^4) = r(1) \times r(1)。$$

如果在等式左边乘以 137 个 $r(7^4)$，要想让左右依然相等，则右边需乘 137 个 $r(1)$，即

$$r(7^4) \cdots r(7^4) = r(1) \cdots r(1)$$

根据高斯同余乘法定理，则有：

$$r(7^4 \cdots 7^4) = r(1 \cdots 1)，$$

即

$$r(7^{548}) = r(1) = 1。$$

有同学可能会问，548 是恰好能被 4 整除，如果问 7^{549} 除以 10 余数是多少，那该怎么办呢？

因为原理上完全相同，所以我们只需在两边继续同时乘以一个 $r(7)$ 即可。

$$r(7^{548}) \times r(7) = r(1) \times r(7)$$
$$r(7^{549}) = r(7) = 7$$

很多初学者喜欢用尾数循环的办法来解决类似的大数求余问题。这不是一个稳妥的办法，有些除数的情况（例如 10）可以用，是因为无论被除数的十位数怎么变化，只需要对应商那个数，余数永远等于个位数。但当出现一些比较特殊的除数时，很容易让整个循环的规律不成立，所以最好的方法还是反复使用高斯同余的乘法定理。

下面让我们再练一题，5^{171} 除以 6 余数为几？

第一步：寻找 5 的多少次方除以 6 余 1。

5^1 除以 6 余 5；5^2 除以 6 余 1。

第二步：利用高斯同余乘法定理解题。

$$r(5^2) = r(1)，$$
$$r(5^2 \cdots 5^2) = r(1 \cdots 1)，$$
$$r(5^{170}) = r(1)。$$

题目问的是 5 的 171 次方，因此等式两边同时乘以 $r(5)$，则有：

$$r(5^{170}) \ r(5) = r(1) \times r(5)。$$

因为 5 除以 6 余 5，所以，5^{171} 除以 6 余 5。

例题 21

What is the remainder when 2^{5500} is divided by 7?

(A) 1 　　　(B) 2 　　　(C) 3 　　　(D) 4 　　　(E) 5

解：

第一步：寻找 2 的多少次方除以 7 余 1。

2^1 除以 7 余 2；2^2 除以 7 余 4；2^3 除以 7 余 1。

第二步：利用高斯同余乘法定理解题。

$$r(2^3) = r(1)，$$

等式两边同时乘以自己，则有：

$$r(2^{5499}) = r(1)，$$
$$r(2^{5499}) \times r(2) = r(1) \times r(2)。$$

因为 2 除以 7 余 2，所以，2^{5500} 除以 7 余 2。

答案为 B。

例题 22

What is the remainder when 3^{24} is divided by 5?

(A) 0　　　　(B) 1　　　　(C) 2　　　　(D) 3　　　　(E) 4

解:

第一步: 寻找 3 的多少次方除以 5 余 1。

3^1 除以 5 余 3;3^2 除以 5 余 4;3^3 除以 5 余 2;3^4 除以 5 余 1。

第二步: 利用高斯同余乘法定理解题。

$$r(3^4) = r(1),$$

等式两边同时乘以自己,则有:

$$r(3^{24}) = r(1)。$$

因此,3 的 24 次方除以 5 余 1。

答案为 B。

例题 23

If n and m are positive integers, what is the remainder when $3^{(4n+2+m)}$ is divided by 10?

(1) $n = 2$

(2) $m = 1$

(A) Statement (1) ALONE is sufficient, but statement (2) alone is not sufficient.

(B) Statement (2) ALONE is sufficient, but statement (1) alone is not sufficient.

(C) BOTH statements TOGETHER are sufficient, but NEITHER statement ALONE is sufficient.

(D) EACH statement ALONE is sufficient.

(E) Statements (1) and (2) TOGETHER are NOT sufficient.

解：

题目问的是，3^{4n+2+m}除以 10 余数为几。

首先找 3 的多少次方除以 10 余 1。

3^1除以 10 余 3；3^2除以 10 余 9；3^3除以 10 余 7；3^4除以 10 余 1。

根据高斯同余乘法定理，在除以 10 的时候，无论 n 等于几，$r(3^{4n})$ 一定是 1。

$$r(3^{4n+2+m}) = r(3^{4n}) \times r(3^2) \times r(3^m) = 1 \times 9 \times r(3^m),$$

因此，只需要知道 3^m 除以 10 余数为几即可。

显然，条件 2 充分，答案为 B。

例题 24

If n is a positive integer, what is the remainder when $3^{8n+3} + 2$ is divided by 5?

(A) 0 　　　　(B) 1 　　　　(C) 2 　　　　(D) 3 　　　　(E) 4

解：

第一步：寻找 3 的多少次方除以 5 余 1。

$$3^1 \text{ 除以 5 余 3；} 3^2 \text{ 除以 5 余 4；} 3^3 \text{ 除以 5 余 2；} 3^4 \text{ 除以 5 余 1。}$$

第二步：利用高斯同余乘法定理解题。

$$r(3^4) = r(1),$$

根据高斯同余乘法定理，在除以 10 的时候，无论 n 等于几，$r(3^{8n})$ 一定是 1，

$$r(3^{8n}) = r(1),$$

因此，

$$r(3^{8n+3}) = r(3^{8n}) \times r(3^3) = 1 \times r(3^3) = 2,$$

根据高斯同余加法定理，可得出：

$$r\left(3^{8n+3}+2\right)=r\left(3^{8n+3}\right)+r(2)=4,$$

答案为 E。

④ 中国剩余定理

中国剩余定理，又叫"孙子定理"，这个名字起源于"韩信点兵"的历史故事。演化到如今的数学题目，就变为：

有一个正整数 n，这个数除以 5 余 1 且除以 7 余 2，问这个数最小是几？

类似这样的考题，我们不必像小学奥数一样用"拼凑法"，而是直接试数即可。

我们先寻找到一系列除以 7 余 2 的数。最小的除以 7 余 2 的数就是 2；第二小的数是 2 + 7；第三小的数是 2 + 7 + 7，以此类推即可。

这件事可以很容易用高斯同余加法定理证明得出，大家可以自己试一试。假设一个数 a 除以 m 余 n，则 $a+km$（k 是任意非负整数）除以 m 依然余 n。

找到这些除以 7 余 2 的数后，只需从最小的数开始判断是否能除以 5 余 1 即可。

2 除以 5 余 2；9 除以 5 余 4；16 除以 5 余 1。

因此，最小的正整数 n 为 16。基于同余定理可知，任意 $16+35k$（k 是任意非负整数）均可除以 5 余 1 且除以 7 余 2。此处，35 刚好是能同时被 5 和 7 整除的最小的数字（5 和 7 的最小公倍数）。

例题 25

When positive integer n is divided by 5, the remainder is 1. When n is divided by 7, the remainder is 3. What is the smallest positive integer k such that $k+n$ is a multiple of 35?

(A) 3   (B) 4   (C) 12   (D) 32   (E) 35

解：

n 除以 5 余 1 且 n 除以 7 余 3，我们可以优先确定 n 的可能性。

先找到一系列除以 7 余 3 的数。最小的除以 7 余 3 的数就是 3；第二小的数是 $3+7$；第三小的数是 $3+7+7$。

再找其中哪个数可以除以 5 余 1。

3 除以 5 余 3；10 除以 5 余 0；17 除以 5 余 2；24 除以 5 余 4；31 除以 5 余 1。

因此，n 可以写为 $31+35k$（k 是任意非负整数）。

题目要求 $k+n$ 是 35 的倍数，所以 $31+k$ 必须是 35 的倍数。k 最小应为 4，答案为 B。

例题 26

When positive integer x is divided by 5, the remainder is 3; and when x is divided by 7, the remainder is 4. When positive integer y is divided by 5, the remainder is 3; and when y is divided by 7, the remainder is 4. If $x > y$, which of the following must be a factor of $x - y$?

(A) 12 　　　　 (B) 15 　　　　 (C) 20

(D) 28 　　　　 (E) 35

解：

根据中国剩余定理，x 和 y 均可以写为 $18+35k$。由于 x 大于 y，所以 $x-y$ 必为：

$$18+35m-18-35n=35(m-n) \quad (m>n)$$

显然，$x-y$ 必然为 35 的倍数。因此，35 一定是 $x-y$ 的因数，答案为 E。

例题 27

When positive integer n is divided by 3, the remainder is 2; and when positive integer t is divided by 5, the remainder is 3. What is the remainder when the product nt is divided by 15?

(1) $n - 2$ is divisible by 5.

(2) t is divisible by 3.

(A) Statement (1) ALONE is sufficient, but statement (2) alone is not sufficient.

(B) Statement (2) ALONE is sufficient, but statement (1) alone is not sufficient.

(C) BOTH statements TOGETHER are sufficient, but NEITHER statement ALONE is sufficient.

(D) EACH statement ALONE is sufficient.

(E) Statements (1) and (2) TOGETHER are NOT sufficient.

解:

题目问的是 $n \times t$ 除以 15 余数为几。

因此，要想知道问题的答案，需要知道 n 和 t 的情况。

条件 1 说，$n-2$ 能被 5 整除。这表明 n 除以 5 余 2。故条件 1 不充分。

条件 2 说，t 可以被 3 整除，这表明 t 除以 3 余 0。故条件 2 不充分。

条件 1 + 条件 2，因为 n 除以 3 余 2，且 n 除以 5 余 2，所以根据中国剩余定理，n 可以写为 $2 + 15k$；同理，t 可以写为 $3 + 15m$。此时 $n \times t$ 除以 15 的余数可以写为：

$$r(n \times t) = r\left[(2 + 15k) \times (3 + 15m)\right] = r(6) + r(30m) + r(45k) + r(225k \times m) = 6 + 0 + 0 + 0 = 6。$$

故条件 1 + 条件 2 充分。

综上，答案为 C。

例题 28

When the positive integer *n* is divided by 25, the remainder is 13. What is the value of *n*?

(1) $n < 100$

(2) When *n* is divided by 20, the remainder is 3.

(A) Statement (1) ALONE is sufficient, but statement (2) alone is not sufficient.

(B) Statement (2) ALONE is sufficient, but statement (1) alone is not sufficient.

(C) BOTH statements TOGETHER are sufficient, but NEITHER statement ALONE is sufficient.

(D) EACH statement ALONE is sufficient.

(E) Statements (1) and (2) TOGETHER are NOT sufficient.

解:

条件 1 说, *n* 小于 100, 小于 100 且除以 25 余 13 的数显然不止一个, 故条件 1 不充分。

条件 2 说, *n* 除以 20 余 3。

根据中国剩余定理可知, 除以 25 余 13 的数有 13, 38, 63 等, 其中, 63 除以 20 余 3。由此可知, $n = 63 + 100k$。无法确定唯一的 *n* 的数值。

条件 1 + 条件 2 可知, *n* 只能等于 63。

综上, 答案为 C。

5 万年历

GMAT 偶尔也会把考生当万年历, 来算闰年, 算日子。实际上这类问题都属于余数问题。

首先是闰年问题, 记住两个计算方式即可。

(1) 普通年能被 4 整除且不能被 100 整除的为闰年。(如 2004 年就是闰年, 1900 年就不是闰年。)

(2) 世纪年能被 400 整除的是闰年。(如 2000 年是闰年, 1900 年就不是闰年。)

其次是推算星期几的问题。例如，假设 1989 年 7 月 1 日是星期三，问 1997 年 7 月 1 日是星期几。类似这样的问题，计算方式为：

普通年一年 365 天；闰年一年 366 天。一周有 7 天。

365 除以 7 余 1，因此，每过一个普通年，同一日期的星期需向后推一天；

366 除以 7 余 2，因此，每过一个闰年，同一日期的星期需向后推两天。

> **例题 29**
>
> June 25, 1982, fell on a Friday. On which day of the week did June 25, 1987, fall?
> (Note: 1984 was a leap year.)
>
> (A) Sunday   (B) Monday   (C) Tuesday
> (D) Wednesday   (E) Thursday
>
> **解：**
>
> 1983，1985，1986，1987 是普通年，同一日期的星期向后推一天；1984 是闰年，同一日期的星期向后推两天，因此，如果 1982 年 6 月 25 日是星期五，则 1987 年 6 月 25 日是星期四（向后加 6 天）。答案为 E。

2.4 ▶ 分数

在分数 $\frac{n}{d}$ 中，n 叫作分子（numerator），d 叫作分母（denominator），分母必然不能为 0。

对于加减法，请先进行通分，之后将分子进行加减，例如：

$$\frac{2}{3} + \frac{1}{6} = \frac{2}{3} \times \frac{2}{2} + \frac{1}{6} = \frac{4}{6} + \frac{1}{6} = \frac{5}{6}。$$

对于乘法，请将分子、分母分别相乘，例如：

$$\frac{2}{3} \times \frac{4}{7} = \frac{2 \times 4}{3 \times 7} = \frac{8}{21}。$$

对于除法，请将除数变为倒数（reciprocal），之后进行乘法运算，例如：

$$\frac{2}{3} \div \frac{4}{7} = \frac{2}{3} \times \frac{7}{4} = \frac{14}{12} = \frac{7}{6} \circ$$

例题 1

Members of a social club met to address 280 newsletters. If they addressed $\frac{1}{4}$ of the newsletters during the first hour and $\frac{2}{5}$ of the remaining newsletters during the second hour, how many newsletters did they address during the second hour?

(A) 28 　　　(B) 42 　　　(C) 63 　　　(D) 84 　　　(E) 112

解：

第 1 个小时处理后剩余的数量为：$280 \times \left(1 - \frac{1}{4}\right) = 210$；第二个小时处理的数量为：

$210 \times \frac{2}{5} = 84$。

综上，答案是 D。

例题 2

$$\cfrac{1}{3 - \cfrac{1}{3 - \cfrac{1}{3 - 1}}} =$$

(A) $\frac{7}{23}$ 　　　(B) $\frac{5}{13}$ 　　　(C) $\frac{2}{3}$ 　　　(D) $\frac{23}{7}$ 　　　(E) $\frac{13}{5}$

解：

一层一层地从里到外计算：

$$\cfrac{1}{3 - \cfrac{1}{3 - \cfrac{1}{3 - 1}}} = \cfrac{1}{3 - \cfrac{1}{3 - \cfrac{1}{2}}} = \cfrac{1}{3 - \cfrac{1}{\cfrac{5}{2}}} = \cfrac{1}{3 - \cfrac{2}{5}} = \cfrac{1}{\cfrac{13}{5}} = \frac{5}{13} \circ$$

综上，答案为 B。

例题 3

$$\frac{4+5^2}{1+2^{-1}} =$$

（A）$\dfrac{14}{5}$ （B）$\dfrac{58}{3}$ （C）$\dfrac{8}{5}$ （D）3 （E）1

解：

将分子进行运算后可得 29；将分母进行运算后可得 $\dfrac{3}{2}$。

综上，答案为 B。

例题 4

If m and n are positive integers, what is the value of $\dfrac{3}{m}+\dfrac{n}{4}$?

（1）$m \times n = 12$

（2）$\dfrac{3}{m}$ is in lowest terms and $\dfrac{n}{4}$ is in lowest terms.

（A）Statement（1）ALONE is sufficient, but statement（2）alone is not sufficient.

（B）Statement（2）ALONE is sufficient, but statement（1）alone is not sufficient.

（C）BOTH statements TOGETHER are sufficient, but NEITHER statement ALONE is sufficient.

（D）EACH statement ALONE is sufficient.

（E）Statements（1）and（2）TOGETHER are NOT sufficient.

解：

$\dfrac{3}{m}+\dfrac{n}{4}=\dfrac{12+m \times n}{4m}$。由此可知，我们只需要知道 m 和 $m \times n$ 的值就可求解。

条件 1 告诉我们 $m \times n$ 的结果，故条件 1 不充分。

条件 2 告诉我们 $\dfrac{3}{m}$ 和 $\dfrac{n}{4}$ 都是最简分数（in lowest terms）。这表明 n 不是 2 的倍数且 m 中没有质因数 3，依然无法知道 m 的值，故条件 2 不充分。

条件 1 + 条件 2，如果 $\dfrac{3}{m}$ 和 $\dfrac{n}{4}$ 都是最简分数，且 $m \times n = 12$，则 m 必为 4，n 必为 3。

故条件 1 + 条件 2 充分。

综上，答案为 C。

例题 5

Of the goose eggs laid at a certain pond, $\frac{2}{3}$ hatched, and $\frac{3}{4}$ of the geese that hatched from those eggs survived the first month. Of the geese that survived the first month, $\frac{3}{5}$ did not survive the first year. If 120 geese survived the first year and if no more than one goose hatched from each egg, how many goose eggs were laid at the pond?

（A）280　　　（B）400　　　（C）540　　　（D）600　　　（E）840

解：

设鹅蛋的总数为 x，则有

$$\frac{2}{3} \times \frac{3}{4} \times \left(1 - \frac{3}{5}\right)x = 120,$$

$$x = 600_{\circ}$$

答案为 D。

例题 6

On a map, $\frac{1}{2}$ inch represents 100 miles. According to this map, how many miles is City X from City Y?

（1）City X is 3 inches from City Y on the map.

（2）Cities X and Y are each 300 miles from City Z.

（A）Statement（1）ALONE is sufficient, but statement（2）alone is not sufficient.

（B）Statement（2）ALONE is sufficient, but statement（1）alone is not sufficient.

（C）BOTH statements TOGETHER are sufficient, but NEITHER statement ALONE is sufficient.

（D）EACH statement ALONE is sufficient.

（E）Statements（1）and（2）TOGETHER are NOT sufficient.

解：

题目问的是城市 X 到城市 Y 的距离是多少。已经给出了换算方式，1inch＝200miles。

条件 1 说，在地图上两者距离为 3 英尺。显然可以通过题干的换算方式进行计算，即 600 英里，故条件 1 充分。

条件 2 中我们无法得知城市 Z 的位置，进而无法知道城市 X 和城市 Y 之间的距离，故条件 2 不充分。

综上，答案为 A。

例题 7

What is the value of $\dfrac{x}{yz}$?

(1) $x = \dfrac{y}{2}$ and $z = \dfrac{2x}{5}$.

(2) $\dfrac{x}{z} = \dfrac{5}{2}$ and $\dfrac{1}{y} = \dfrac{1}{10}$.

(A) Statement (1) ALONE is sufficient, but statement (2) alone is not sufficient.

(B) Statement (2) ALONE is sufficient, but statement (1) alone is not sufficient.

(C) BOTH statements TOGETHER are sufficient, but NEITHER statement ALONE is sufficient.

(D) EACH statement ALONE is sufficient.

(E) Statements (1) and (2) TOGETHER are NOT sufficient.

解：

根据条件 1，我们可以将 $\dfrac{x}{yz}$ 的分子和分母均用 y 来表示。显然分母中会出现 y^2，而分子中只有一个 y，无法约分，即无法确定 $\dfrac{x}{yz}$ 的值。故条件 1 不充分。

根据条件 2，$\dfrac{x}{yz} = \dfrac{x}{z} \times \dfrac{1}{y} = \dfrac{5}{2} \times \dfrac{1}{10} = \dfrac{1}{4}$。故条件 2 充分。

综上，答案为 B。

2.5 ▶ 小数

我们先来认识一下数字中各位的英文叫法，例如 7,654.321。

Thousands		Hundreds	Tens	Ones or units		Tenths	Hundredths	Thousandths
7	,	6	5	4	.	3	2	1

有些小数会表达为如下形式：

$$0.321 = \frac{3}{10} + \frac{2}{100} + \frac{1}{1,000} = \frac{321}{1,000},$$

$$0.0321 = \frac{0}{10} + \frac{3}{100} + \frac{2}{1,000} + \frac{1}{10,000} = \frac{321}{10,000},$$

$$1.56 = 1 + \frac{5}{10} + \frac{6}{100} = \frac{156}{100}。$$

考题中经常会出现科学计数法。这是一种记数的方法，把一个数表示成 a 与 10 的 n 次幂相乘的形式（$1 \leq |a| < 10$，n 为整数），例如：0.0231 可以写作 2.31×10^{-2}。当然，不仅是小数可以用科学计数法来计数，整数也可以用科学计数法来表达，例如：19,971,400,000,000 可以表达为 1.99714×10^{13}。考试中常常让我们判断科学计数法背后的数值。

例题 1

If $10^{50} - 74$ is written as an integer in base 10 notation, what is the sum of the digits in that integer?

(A) 424　　　(B) 433　　　(C) 440　　　(D) 449　　　(E) 467

解：

base 10 notation 的意思是十进制的整数。10 的 50 次方表示的是 1 后面有 50 个 0，这个数字减去 74 后，形如：

$$\begin{array}{r} 10\cdots000 \\ -\qquad 74 \\ \hline 9\cdots926 \end{array}$$

通过列出标准的竖式，很容易数清楚位数。最后两位变为 2 和 6，之前的所有 0 均会变为 9。因此，一共有 48 个 9、一个 2 以及一个 6。它们的和为 440，答案为 C。

定量推理简介
第一章

算数
第二章

代数
第三章

几何
第四章

文字问题
第五章

例题 2

A positive integer is divisible by 9 if and only if the sum of its digits is divisible by 9. If n is a positive integer, for which of the following values of k is $25 \times 10^{n} + k \times 10^{2n}$ divisible by 9?

(A) 9　　　　(B) 16　　　　(C) 23　　　　(D) 35　　　　(E) 47

解：

无论 n 等于几，它们都只会在 25 和 k 后面添加 0。因此，n 等于几并不影响 $25 \times 10^{n} + k \times 10^{2n}$ 各个数位的数值之和。因此，只需要将五个选项分别代入，检查几个非 0 数位相加是否能被 9 整除即可，只有 $2+5+4+7=18$，可以被 9 整除。答案为 E。

例题 3

If $1 < d < 2$, is the tenths digit of the decimal representation of d equal to 9?

(1) $d + 0.01 < 2$

(2) $d + 0.05 > 2$

(A) Statement (1) ALONE is sufficient, but statement (2) alone is not sufficient.

(B) Statement (2) ALONE is sufficient, but statement (1) alone is not sufficient.

(C) BOTH statements TOGETHER are sufficient, but NEITHER statement ALONE is sufficient.

(D) EACH statement ALONE is sufficient.

(E) Statements (1) and (2) TOGETHER are NOT sufficient.

解：

题目问的是 d 的十分位是否等于 9。已经给出 d 的取值范围是 $1 \sim 2$。

条件 1 说，$d + 0.01 < 2$。移项可得，$d < 1.99$。显然，小于 1.99 且大于 1 的数很多，不是每一个数的十分位都是 9（例如 1.89），故条件 1 不充分。

条件 2 说，$d + 0.05 > 2$。移项可得，$d > 1.95$。所有大于 1.95 且小于 2 的数的十分位均为 9，故条件 2 充分。

综上，答案为 B。

> **例题 4**
>
> If k is an integer and $(0.0025)(0.025)(0.00025) \times 10^k$ is an integer, what is the least possible value of k?
>
> (A) −12 (B) −6 (C) 0 (D) 6 (E) 12
>
> 解:
>
> 因为 25 的末尾数字是 5，所以要想让题干中的数字为整数，务必要让 0.025，0.0025，0.00025 这些数字变成 25。
>
> $$0.0025 = 25 \times 10^{-4},$$
> $$0.025 = 25 \times 10^{-3},$$
> $$0.00025 = 25 \times 10^{-5}。$$
>
> 因此，$k = 12$。
>
> 综上，答案为 E。

请注意，在对诸如 0.025 这样的数字进行科学计数法时，可以用"挪动小数点"法。

例如:

$$0.0025 = 0.025 \times 10^{-1} = 0.25 \times 10^{-2} = 2.5 \times 10^{-3}。$$

诀窍是，挪动几位，就在后面乘以 10 的负几次方。

2.5.1 ▶ 四舍五入

GMAT 中有很多方式来表达四舍五入:

round to

to the nearest

round up to

Round down to

前两个就是四舍五入的意思，round up to 表示只入不舍，round down to 表示只舍不入。例如，7,654.321，round to tenths，则可以写为 7,654.3。

考题中经常会给出四舍五入后的结果，让我们判断实际的取值范围。

例题 4

On a recent trip, Cindy drove her car 290 miles, rounded to the nearest 10 miles, and used 12 gallons of gasoline, rounded to the nearest gallon. The actual number of miles per gallon that Cindy's car got on this trip must have been between

(A) $\dfrac{290}{12.5}$ and $\dfrac{290}{11.5}$.　　　(B) $\dfrac{295}{12}$ and $\dfrac{285}{11.5}$.　　　(C) $\dfrac{285}{12}$ and $\dfrac{295}{12}$.

(D) $\dfrac{285}{12.5}$ and $\dfrac{295}{11.5}$.　　(E) $\dfrac{295}{12.5}$ and $\dfrac{285}{11.5}$.

解：

路程四舍五入到十位数后是 290 公里，说明实际数字应是 285 ~ 295（不含 295）；耗油量四舍五入到个位数后是 12 加仑，说明实际数字应该是 11.5 ~ 12.5（不含 12.5）。两个数比值最大为 $\dfrac{295}{11.5}$；最小为 $\dfrac{285}{12.5}$，故答案为 D。

例题 5

What is the value of the integer x?

(1) x rounded to the nearest hundred is 7,200.

(2) The hundreds digit of x is 2.

(A) Statement (1) ALONE is sufficient, but statement (2) alone is not sufficient.

(B) Statement (2) ALONE is sufficient, but statement (1) alone is not sufficient.

(C) BOTH statements TOGETHER are sufficient, but NEITHER statement ALONE is sufficient.

(D) EACH statement ALONE is sufficient.

(E) Statements (1) and (2) TOGETHER are NOT sufficient.

解：

条件 1 说，四舍五入至百位，$x = 7200$。显然，十位和个位不能确定，故条件 1 不充分。

条件 2 说，百位数字是 2，依然不能确定十位和个位数字，故条件 2 不充分。

两个条件同时满足时，十位和个位数字仍不能确定。

综上，答案为 E。

例题 6

What is the result when x is rounded to the nearest hundredth?

(1) When x is rounded to the nearest thousandth the result is 0.455.

(2) The thousandths digit of x is 5.

(A) Statement (1) ALONE is sufficient, but statement (2) alone is not sufficient.

(B) Statement (2) ALONE is sufficient, but statement (1) alone is not sufficient.

(C) BOTH statements TOGETHER are sufficient, but NEITHER statement ALONE is sufficient.

(D) EACH statement ALONE is sufficient.

(E) Statements (1) and (2) TOGETHER are NOT sufficient.

解：

条件 1 说，x 四舍五入到千分位的结果是 0.455。这表明，x 可以是 0.4551、0.4552 等，此时 x 四舍五入到百分位的结果为 0.46；x 也可以是 0.4549、0.4548 等，此时 x 四舍五入到百分位的结果为 0.45。因此无法确定 x 的值，故条件 1 不充分。

条件 2 说，x 的千分位是 5，依然不能确定 x 四舍五入至百分位的值，故条件 2 不充分。

两个条件同时满足时，由于 x 的千分位是 5，且 x 四舍五入到千分位的结果是 0.455，此时必然能确定 x 四舍五入至百分位的结果为 0.46，故条件 1 + 条件 2 充分。

综上，答案为 C。

例题 7

A country's per capita national debt is its national debt divided by its population. Is the per capita national debt of Country G within $5 of $500?

(1) Country G's national debt to the nearest $1,000,000,000 is $43,000,000,000

(2) Country G's population to the nearest 1,000,000 is 86,000,000

(A) Statement (1) ALONE is sufficient, but statement (2) alone is not sufficient.

(B) Statement (2) ALONE is sufficient, but statement (1) alone is not sufficient.

(C) BOTH statements TOGETHER are sufficient, but NEITHER statement ALONE is sufficient.

(D) EACH statement ALONE is sufficient.

(E) Statements (1) and (2) TOGETHER are NOT sufficient.

解：

题目问的是债务除以人口是否在 495～505 之间。

条件 1 说，债务四舍五入到十亿位为 43,000,000,000，即 42,500,000,000 ≤ 债务 ≤ 43,400,000,000。因为不知道人口数量，故条件 1 不充分。

条件 2 说，人口四舍五入到百万位为 86,000,000，则 85,500,000 ≤ 人口 ≤ 86,400,000。因为不知道债务数额，故条件 2 不充分。

两个条件同时成立时，$\dfrac{42,500,000,000}{86,400,000}$ 得到的是 "$\dfrac{债务}{人口}$" 的下限；$\dfrac{43,400,000,000}{85,500,000}$ 得到的是 "$\dfrac{债务}{人口}$" 的上限，显然其范围大于 495～505，故条件 1 + 条件 2 依然不充分。

综上，答案为 E。

例题 8

If $d = 2.0453$ and $d*$ is the decimal obtained by rounding d to the nearest hundredth, what is the value of $d* - d$?

(A) -0.0053 (B) -0.0003 (C) 0.0007 (D) 0.0047 (E) 0.0153

解：

$d*$ 是四舍五入到 d 的百分位，所以 $d* = 2.05$。

$$d* - d = 2.05 - 2.0453 = 0.0047$$

综上，答案为 D。

2.5.2 ▶ 有限小数

所谓有限小数（terminating decimal），指的是那些长度有限的小数，而不是那些循环或不循环的无限小数，如著名的圆周率就是无限不循环小数，并非有限小数。

判断一个分数是否是有限小数非常简单，如果能确保这个分数是最简分数，则对其分母进行质因数分解。分解之后发现分母中只含有 2 或 5 这两个质因数，则该分数必为有限小数。如果分母中还含有其他的质因数，则该分数必然不是有限小数。

例题 9

Which of the following fractions has a decimal equivalent that is a terminating decimal?

(A) $\dfrac{10}{189}$　　(B) $\dfrac{15}{196}$　　(C) $\dfrac{16}{225}$　　(D) $\dfrac{25}{144}$　　(E) $\dfrac{39}{128}$

解：

五个选项均已是最简分数，直接对分母进行质因数分解即可。

$$189 = 3^3 \times 7 \text{；} 196 = 2^2 \times 7^2 \text{；} 225 = 3^2 \times 5^2 \text{；} 144 = 2^4 \times 3^2 \text{；} 128 = 2^7 \text{。}$$

显然，只有选项 E 的分母的质因数只有 2，答案为 E。

2.6 ▶ 比例和比率

在解涉及比例和比率的题目时，最大的障碍是很容易读错题意。请注意以下四种表达方式：

What percent of A is B? 这个问题问的是 $\dfrac{B}{A}$。

What is the ratio of *A* to *B*? 这个问题问的是 *A*:*B*。

A is three times as many as *B*; There is three times as many *A* as *B*. 这两句话的意思都是 $A = 3B$。

A is three times more than *B*. 这句话的意思是 $A = 4B$。

另外，GMAT 考题很喜欢问增长或下降了百分之多少 $\left(\dfrac{\text{percent increase}}{\text{percent decrease}}\right)$。这类的问题请记住一个表达式：

$$增长率（下降率）= \frac{|\text{现值} - \text{原值}|}{\text{原值}}$$

例题 1

The population of City X is 50 percent of the population of City Y. The population of City X is what percent of the total population of City X and City Y?

（A）25%   （B）$33\frac{1}{3}$%   （C）40%   （D）50%   （E）$66\frac{2}{3}$%

解：

依题意，X = 0.5Y，

题目问的是 $\dfrac{X}{X + Y}$：

$$\frac{X}{X + Y} = \frac{0.5}{1.5} = \frac{1}{3},$$

综上，答案为 B。

例题 2

On July 1 of last year, the total number of employees at Company E was decreased by 10 percent. Without any change in the salaries of the remaining employees, the average (arithmetic mean) employee salary was 10 percent more after the decrease in number of employees than before the decrease. The total of the combined salaries of all of the employees at Company E after July 1 last year was what percent of that before July 1 last year?

(A) 90%　　(B) 99%　　(C) 100%　　(D) 101%　　(E) 110%

解:

假设 7 月 1 日前的员工数为 E，则 7 月 1 日后的员工数为 $0.9E$；假设 7 月 1 日前的平均薪水为 S，则 7 月 1 日后的平均薪水为 $1.1S$。7 月 1 日前的总薪水为 $E \times S$；7 月 1 日后的总薪水为 $0.9E \times 1.1S$。

7 月 1 日后：7 月 1 日前 $= \dfrac{0.9E \times 1.1S}{E \times S} = 0.99 = 99\%$。答案为 B。

例题 3

A grocer has 400 pounds of coffee in stock, 20 percent of which is decaffeinated. If the grocer buys another 100 pounds of coffee of which 60 percent is decaffeinated, what percent, by weight, of the grocer's stock of coffee is decaffeinated?

(A) 28%　　(B) 30%　　(C) 32%　　(D) 34%　　(E) 40%

解:

先计算出 decaffeinated coffee 的含量:

$$400 \times 20\% + 100 \times 60\% = 140。$$

现在总的咖啡量:

$$400 + 100 = 500。$$

占比为:

$$\frac{140}{500} = 28\%。$$

答案为 A。

例题 4

At the opening of a trading day at a certain stock exchange, the price per share of stock K was $8. If the price per share of stock K was $9 at the closing of the day, what was the percent increase in the price per share of stock K for that day?

（A）1.4%　　（B）5.9%　　（C）11.1%　　（D）12.5%　　（E）23.6%

解：

现值是 $9，原值是 $8，增长率 $= \dfrac{现值 - 原值}{原值} = \dfrac{9-8}{8} = 0.125 = 12.5\%$。答案为 D。

例题 5

If 65 percent of a certain firm's employees are full-time and if there are 5,100 more full-time employees than part-time employees, how many employees does the firm have?

（A）8,250　　（B）10,200　　（C）11,050　　（D）16,500　　（E）17,000

解：

可以设兼职员工数量为 x，则全职员工数量为：$x + 5100$。

依题意，可以列方程为：

$$0.65(x + x + 5100) = x + 5100,$$

$$x = 5950。$$

全体员工数为：

$$x + x + 5100 = 17000。$$

综上，答案为 E。

例题 6

Last Sunday a certain store sold copies of Newspaper A for $1.00 each and copies of Newspaper B for $1.25 each, and the store sold no other newspapers that day. If r percent of the store's revenues from newspaper sales was from Newspaper A and if p percent of the newspapers that the store sold were copies of newspaper A, which of the following expresses r in terms of p?

（A）$\dfrac{100p}{125 - p}$　　（B）$\dfrac{150p}{250 - p}$　　（C）$\dfrac{300p}{375 - p}$　　（D）$\dfrac{400p}{500 - p}$　　（E）$\dfrac{500p}{625 - p}$

解：

依题意，卖出的 A 报纸的份数比例是 p，那么卖出的 B 报纸的份数比例就是 $100 - p$。A 报纸和 B 报纸的价格比是 $\dfrac{1}{1.25} = \dfrac{4}{5}$。

因此，若卖出的 A 报纸的收入是 $4p$，则卖出的 B 报纸的收入就是 $5(100 - p)$，

然后，卖出的 A 报纸的收入占所有收入的百分比是 r，所以

$$\frac{4p}{500 - 5p + 4p} = \frac{r}{100},$$

整理后得出：

$$r = \frac{400p}{500 - p},$$

综上，答案为 D。

例题 7

Of the total amount that Jill spent on a shopping trip, excluding taxes, she spent 50 percent on clothing, 20 percent on food, and 30 percent on other items. If Jill paid a 4 percent tax on the clothing, no tax on the food, and an 8 percent tax on all other items, then the total tax that she paid was what percent of the total amount that she spent, excluding taxes?

(A) 2.8%   (B) 3.6%   (C) 4.4%   (D) 5.2%   (E) 6.0%

解：

首先假设 Jill 一共花费 x 元；买衣服不含税花费 $50\% \cdot x$，食品不含税花费是 $20\% \cdot x$，买其他东西花费 $30\% \cdot x$。买衣服的税是 $50\% \cdot x \cdot 4\%$，买其他东西的税是 $30\% \cdot x \cdot 8\%$。所求的比例为：

$$(50\% \cdot x \cdot 4\% + 30\% \cdot x \cdot 8\%) x = 4.4\%。$$

答案为 C。

例题 8

When a certain stretch of highway was rebuilt and straightened, the distance along stretch was decreased by 20 percent and the speed limit was increased by 25 percent. By what percent was the driving time along this stretch reduced for a person who always drives at the speed limit?

(A) 16%   (B) 36%   (C) $37\frac{1}{2}$%   (D) 45%   (E) $50\frac{1}{4}$%

解:

设改动前的长度为 d,改动前的限速为 s。则改动后的长度为 $0.8d$,改动后的限速为 $1.25s$。

如果车辆一直是最高速度,则它改动前的通过时间为 $\dfrac{d}{s}$;改动后的通过时间为 $\dfrac{0.8d}{1.25s}$。

$$增长率 = \frac{现值 - 原值}{原值} = \frac{\dfrac{0.8d}{1.25s} - \dfrac{d}{s}}{\dfrac{d}{s}} = -36\%。$$

增长率为 -36% 就意味着减少了 36%。

因此,答案为 B。

例题 9

A certain financial institution reported that its assets totaled \$ 2, 377, 366. 30 on a certain day. Of this amount, \$ 31, 724. 54 was held in cash. Approximately what percent of the reported assets was held in cash on that day?

(A) 0.00013%   (B) 0.0013%   (C) 0.013%   (D) 0.13%   (E) 1.3%

解:

这道题主要考查的是估算问题。题目问的是 $\dfrac{31724.54}{2377366.30}$ 约等于多少。我们可以把分子、分母同时缩小 10000 倍,则问题中的比值约等于 $\dfrac{3.2}{237}$。由此可知,两者的比值大约为 0.01,即 1%。综上,答案为 E。

例题 10

A grocer has 400 pounds of coffee in stock, 20 percent of which is decaffeinated. If the grocer buys another 100 pounds of coffee of which 60 percent is decaffeinated, what percent, by weight, of the grocer's stock of coffee is decaffeinated?

（A）28%　　　（B）30%　　　（C）32%　　　（D）34%　　　（E）40%

解：

依题意，原来不含咖啡因的咖啡库存量为 $400 \times 0.2 = 80$。新增的不含咖啡因的咖啡量为 $100 \times 0.6 = 60$。

则现在不含咖啡因的咖啡库存占比为：$\dfrac{80+60}{500} = 28\%$。答案为 A。

例题 11

If $5x = 2y$ and $xy \neq 0$, then x is what percent of y?

（A）2%　　　（B）4%　　　（C）10%　　　（D）40%　　　（E）50%

解：

这道题主要需要看懂题目，题目问的是 x 是 y 的百分之多少。

$$x = \left(\frac{2}{5}\right)y = 40\%\, y,$$

答案为 D。

例题 12

A rectangular solid has length, width, and height of Lcm, Wcm, and Hcm, respectively. If these dimensions are increased by $x\%$, $y\%$ and $z\%$, respectively, what is the percentage increase in the total surface area of the solid?

（1）L, W, and H are in the ratios of 5:3:4.

（2）$x = 5$, $y = 10$, $z = 20$

（A）Statement（1）ALONE is sufficient, but statement（2）alone is not sufficient.

（B）Statement（2）ALONE is sufficient, but statement（1）alone is not sufficient.

（C）BOTH statements TOGETHER are sufficient, but NEITHER statement ALONE is sufficient.

（D）EACH statement ALONE is sufficient.

（E）Statements（1）and（2）TOGETHER are NOT sufficient.

解：

题目问的是长方体的表面积。长方体表面积＝长×宽×2＋宽×高×2＋长×高×2。

条件1只知道长、宽、高的比值，不知道增长的 x, y, z 的情况，故条件1不充分。

条件2只知道 x, y, z 的情况，不知道长、宽、高的比值，故条件2不充分。

两个条件同时成立时，由于原长方体长、宽、高的比值和增长率的比值都已知，必然能得到表面积增加的百分比。故条件1＋条件2充分。

综上，答案为C。

例题 13

Town X has 50, 000 residents, some of whom were born in Town X. What percent of the residents of Town X were born in Town X?

（1）Of the male residents of Town X, 40 percent were not born in Town X.

（2）Of the female residents of Town X, 60 percent were born in Town X.

（A）Statement（1）ALONE is sufficient, but statement（2）alone is not sufficient.

（B）Statement（2）ALONE is sufficient, but statement（1）alone is not sufficient.

（C）BOTH statements TOGETHER are sufficient, but NEITHER statement ALONE is sufficient.

（D）EACH statement ALONE is sufficient.

（E）Statements（1）and（2）TOGETHER are NOT sufficient.

解：

条件1说，40%的男性不是出生于X镇的。这表示，有60%的男性是出生于X镇的。由于不知道女性的情况，故条件1不充分。

条件2说，60%的女性是出生于X镇的。由于不知道男性的情况，故条件2不充分。

两个条件同时成立时，男女性的出生情况皆已知，故条件1＋条件2充分。

综上，答案为C。

例题 14

On a certain transatlantic crossing, 20 percent of a ship's passengers held round-trip tickets and also took their cars aboard the ship. If 60 percent of the passengers with round-trip tickets did not take their cars aboard the ship, what percent of the ship's passengers held round-trip tickets?

(A) $33\frac{1}{3}\%$　　　(B) 40%　　　(C) 50%　　　(D) 60%　　　(E) $66\frac{2}{3}\%$

解:

依题意,我们知道买了往返票的人有40%带了汽车上船。也就是说,占总人数20%的有往返票又带车的人占据了所有买往返票的人的40%。设既买了往返票又带汽车的人为 x,则:

$$\frac{x}{总人数} = 20\%,$$

$$\frac{x}{买往返票的人数} = 40\%,$$

两式联立则有:

$$\frac{买往返票的人数}{总人数} = \frac{20}{40} = \frac{1}{2} = 50\%,$$

答案为 C。

2.7 幂运算和根运算

2.7.1 幂运算

幂运算(exponentiation)的基本规则,主要需记住一句话:

同底数幂相乘,底数不变,指数相加。同底数幂相除,底数不变,指数相减。幂的乘方,底数不变,指数相乘。

用公式表达为:

(1) 同底数幂的乘法: $a^m \cdot a^n \cdot a^p = a^{m+n+p}$ (m, n, p 都是正整数)。

(2) 幂的乘方 $(a^m)^n = a^{mn}$,与积的乘方 $(ab)^n = a^n b^n$。

(3) 同底数幂的除法:

①同底数幂的除法：$a^m \div a^n = a^{m-n}$（$a \neq 0$，m，n 均为正整数，并且 $m > n$）；

②零指数：$a^0 = 1$（$a \neq 0$）；

③负整数指数幂：$a^{-p} = \dfrac{1}{a^p}$（$a \neq 0$，p 是正整数），当 $a = 0$ 时没有意义。

Which of the following is equal to $\dfrac{2^{12} - 2^6}{2^6 - 2^3}$?

（A）$2^6 + 2^3$　　　（B）$2^6 - 2^3$　　　（C）2^9　　　（D）2^3　　　（E）2

解：

同底数幂不能直接加减，先把分子变为：$(2^6 + 2^3)$ $(2^6 - 2^3)$。与分母约分后，则变为：$2^6 + 2^3$，答案为 A。

例题 2

Which of the following is equal to $5^{17} \times 4^9$?

（A）2×10^{13}　　　　　（B）2×10^{17}　　　　　（C）2×10^{20}

（D）2×10^{26}　　　　　（E）2×10^{36}

解：

本题可以直接通过 5 的个数来计算。因为，10 里一定有一个 5，所以 5 的个数必然等于 10 的个数。因此，答案为 B。

例题 3

If $2^x + 2^x + 2^x + 2^x = 2^n$, what is x in terms of n?

（A）$\dfrac{n}{4}$　　　（B）$4n$　　　（C）$2n$　　　（D）$n - 2$　　　（E）$n + 2$

解：

$$2^x + 2^x + 2^x + 2^x = 4 \times 2^x = 2^2 \times 2^x = 2^{x+2}。$$

因此，$n = x + 2$；$x = n - 2$。答案为 D。

例题 4

What is the value of x?

(1) $4^{x+1} + 4^x = 320$

(2) $x^2 = 9$

(A) Statement (1) ALONE is sufficient, but statement (2) alone is not sufficient.

(B) Statement (2) ALONE is sufficient, but statement (1) alone is not sufficient.

(C) BOTH statements TOGETHER are sufficient, but NEITHER statement ALONE is sufficient.

(D) EACH statement ALONE is sufficient.

(E) Statements (1) and (2) TOGETHER are NOT sufficient.

解：

条件 1，提取公因式，则有：

$$4^x \times (4+1) = 320,$$

$$4^x = \frac{320}{5} = 64,$$

解得 $x = 3$，故条件 1 充分。

条件 2，$x^2 = 9$ 时，x 可以是正数，也可以是负数，即 x 可能等于 3，也可能等于 -3，故条件 2 不充分。

综上，答案为 A。

例题 5

If $x \neq 1$, is y equal to $x + 1$?

(1) $\dfrac{y-2}{x-1} = 1$

(2) $y^2 = (x+1)^2$

(A) Statement (1) ALONE is sufficient, but statement (2) alone is not sufficient.

(B) Statement (2) ALONE is sufficient, but statement (1) alone is not sufficient.

(C) BOTH statements TOGETHER are sufficient, but NEITHER statement ALONE is sufficient.

（D）EACH statement ALONE is sufficient.

（E）Statements（1）and（2）TOGETHER are NOT sufficient.

解：

条件1，等式两边同时乘 $x-1$，则有：

$$y-2=x-1,$$
$$y=x+1。$$

故条件1充分。

条件2，y 可能等于 $x+1$，也可能等于 $-(x+1)$，因此不能确定 y 是否等于 $x+1$，故条件2不充分。

综上，答案为 A。

例题 6

What is the value of x?

（1）$x^4+x^2+1=\dfrac{1}{x^4+x^2+1}$

（2）$x^3+x^2=0$

（A）Statement（1）ALONE is sufficient, but statement（2）alone is not sufficient.

（B）Statement（2）ALONE is sufficient, but statement（1）alone is not sufficient.

（C）BOTH statements TOGETHER are sufficient, but NEITHER statement ALONE is sufficient.

（D）EACH statement ALONE is sufficient.

（E）Statements（1）and（2）TOGETHER are NOT sufficient.

解：

条件1，只有1和 -1 这两个数字能保证和它的倒数相等。x^4 和 x^2 必然为非负数，所以 $x^4+x^2+1=1$。可以确定 $x=0$。故条件1充分。

条件2，$x^3+x^2=x^2(x+1)=0$，

解得 $x=-1$ 或 0，故条件2不充分。

综上，答案为 A。

例题 7

If $y = 2^{x+1}$, what is the value of $y - x$?

(1) $2^{2x+2} = 64$

(2) $y = 2^{2x-1}$

(A) Statement (1) ALONE is sufficient, but statement (2) alone is not sufficient.

(B) Statement (2) ALONE is sufficient, but statement (1) alone is not sufficient.

(C) BOTH statements TOGETHER are sufficient, but NEITHER statement ALONE is sufficient.

(D) EACH statement ALONE is sufficient.

(E) Statements (1) and (2) TOGETHER are NOT sufficient.

解：

条件 1，$2^{2x+2} = 64 = 2^6$，即

$$2x + 2 = 6,$$

解得：

$$x = 2。$$

当 x 等于 2 时，$y = 8$，$y - x = 6$，故条件 1 充分。

条件 2，$y = 2^{x+1} = y = 2^{2x-1}$，即

$$x + 1 = 2x - 1,$$
$$x = 2。$$

当 x 等于 2 时，$y = 8$，$y - x = 6$，故条件 2 充分。

综上，答案为 D。

例题 8

What is the largest integer n such that $\frac{1}{2^n} > 0.01$?

(A) 5　　　　(B) 6　　　　(C) 7　　　　(D) 10　　　　(E) 51

定量推理简介
第一章

算数
第二章

代数
第三章

几何
第四章

文字问题
第五章

解:

$\dfrac{1}{2^n} > 0.01$,整理得出:

$$2^n < 100,$$
$$2^7 = 128,$$
$$2^6 = 64。$$

所以正整数 $n < 7$。则 n 最大为 6。

综上,答案为 B。

2.7.2 ▶ 根运算

根运算(roots)是一种含有开放运算的代数式,即含有根号的表达式。按根指数是偶数还是奇数,根式分别称为偶次根式或奇次根式。根式的性质如下:

① $\sqrt{a^2} = |a| = \begin{cases} a & (a \geqslant 0), \\ -a & (a < 0); \end{cases}$

② $\sqrt[n]{a^n} = \begin{cases} a, & (n \text{ is an odd number}), \\ |a| = \begin{cases} a & (a \geqslant 0), \\ -a & (a < 0), \end{cases} & (n \text{ is an even number}); \end{cases}$

③ $\left(\sqrt[n]{a}\right)^n = a \ (n \in Z)$;

④ $\sqrt[n]{0} = 0$。

例题 9 ▷

If $\left(2 - \sqrt{5}\right) x = -1$, then $x =$

(A) $2 + \sqrt{5}$ (B) $1 + \dfrac{\sqrt{5}}{2}$ (C) $1 - \dfrac{\sqrt{5}}{2}$

(D) $2 - \sqrt{5}$ (E) $-2 - \sqrt{5}$

解：

$$x = \frac{-1}{2 - \sqrt{5}},$$

$$= \frac{1}{\sqrt{5} - 2} = \frac{\sqrt{5} + 2}{(\sqrt{5} - 2)(\sqrt{5} + 2)},$$

$$= \sqrt{5} + 2。$$

故答案为 A。

例题 10

In the formula $W = \frac{P}{\sqrt[t]{V}}$, integers P and t are positive constants. If $W = 2$ when $V = 1$ and

if $W = \frac{1}{2}$ when $V = 64$, then $t =$

(A) 1　　　　(B) 2　　　　(C) 3　　　　(D) 4　　　　(E) 16

解：

将 $V = 1$ 代入公式，由于 1 开多少次方均为 1，所以 $P = 2$。再将 $W = \frac{1}{2}$ 和 $V = 64$ 代入题目，则有

$$t = 3\,(4^3 = 64)。$$

答案为 C。

2.8 ▶ 统计

统计学的概念可谓是数学中最多的，但是大部分的概念都非常简单。

平均数（average/mean）：是指在一组数据中所有数据之和再除以这组数据的个数。

中位数（median）：对于一组有限个数的数据来说，它们的中位数是这样的一种数：这组数据里的一半数据比它大，而另外一半数据比它小。计算有限个数的数据的中位数的方法是：把所有的同类数据按照大小的顺序排列。如果数据的个数是奇数，则中间那个数就是这群数据的中位数；如果数据的个数是偶数，则中间那两个数的平均数就是这群数据的中位数。

众数（mode）：一组数据里出现频次最高的数。如果这种数多于 1 个，那么这些数字都是众数。

极差（range）：一组数据的极大值减去极小值。

方差（variance）：是各个数据与平均数之差的平方的和的平均数。计算公式为：

$$s^2 = \frac{1}{n}\left[(x_1 - x)^2 + (x_2 - x)^2 + \cdots + (x_n - x)^2\right]。$$

其中，x 表示一组数据的平均数，n 表示这组数据的数量，x_i 表示每个数据，而 s^2 就是这组数据的方差。

标准差（standard deviation）：方差的算术平方根。

从物理意义上来说，标准差和方差均反映了一组数据的离散程度。说得直白些，就是反映了一组数据是否波动剧烈。

标准正态分布（standard normal distribution）：

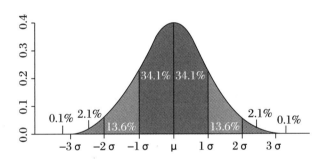

深色区域是距平均值小于一个标准差之内的数值范围。在标准正态分布中，此范围的数值所占比率为全部数值的 68%，根据正态分布，两个标准差之内的比率合起来为 95%；三个标准差之内的比率合起来为 99%。

例题 1

The number line shown contains three points R, S, and T, whose coordinates have absolute values r, s, and t, respectively. Which of the following equals the average (arithmetic mean) of the coordinates of the points R, S, and T?

(A) s　　　　　(B) $s+t-r$　　　　　(C) $\dfrac{r-s-t}{3}$

(D) $\dfrac{r+s+t}{3}$　　　　　(E) $\dfrac{s+t-r}{3}$

解：

本题需仔细读懂题意。r，s，t 是数轴上三个点的绝对值。因为 R 在 0 点左边，因此 R 的真正数值必然是 $-r$。算数平均数应为 $\dfrac{s+t-r}{3}$，答案为 E。

例题 2

The average (arithmetic mean) of the 5 positive integers k, m, r, s and t is 16, and $k<m<r<s<t$. If t is 40, what is the greatest possible value of the median of the 5 integers?

(A) 16　　　　(B) 18　　　　(C) 19　　　　(D) 20　　　　(E) 22

解：

本题有一定难度。因为五个数的平均数为 16，所以五个数的和必然为 $16\times5=80$。又因为 $t=40$，所以 $k+m+r+s=40$。题目的要求是让中位数尽量大。由于 $k<m<r<s$，所以 r 是中位数。想让 r 尽量大，必然是让 k 和 m 尽量小，s 比 r 大得尽量少，这里取 $k=1$，$m=2$，$s=r+1$。（不能取负数和零，因为题干已经规定这五个数均为正整数）。

若 $1+2+r+r+1=40$，则解出 r 的值为 18，满足题目要求，答案为 B。

例题 3

List K consists of 12 consecutive integers. If -4 is the least integer in list K, what is the range of the positive integers in list K?

(A) 5　　　　(B) 6　　　　(C) 7　　　　(D) 11　　　　(E) 12

解:

数列 K 包含 12 个连续整数，最小的是 −4，则最大的应该是 7，注意题干问的是正整数的 range，1~7 的 range 应该是 6，答案是 B。

例题 4

70, 75, 80, 85, 90, 105, 105, 130, 130, 130

The list shown consists of the times, in seconds, that it took each of 10 schoolchildren to run a distance of 400 meters. If the standard deviation of the 10 running times is 22.4 seconds, rounded to the nearest tenth of a second, how many of the 10 running times are more than 1 standard deviation below the mean of the 10 running times?

(A) One　　　(B) Two　　　(C) Three　　　(D) Four　　　(E) Five

解:

因为数学考试中不提供计算器，所以大部分涉及标准差的考题都不会让我们真的去算标准差。题目问的是，有哪几个数是小于平均数又超出一个标准差的。我们先计算平均数，这组数据的平均数为：

$$\text{mean} = \frac{70 + 75 + 80 + 85 + 90 + 105 + 105 + 130 + 130 + 130}{10} = 100$$

因为 $100 - 22.4 = 77.6$，所以题干要求的小于平均数又超出一个标准差的数只有 70 和 75，故答案为 B。

例题 5

A certain characteristic in a large population has a distribution that is symmetric about the mean m. If 68 percent of the distribution lies within one standard deviation d of the mean, what percent of the distribution is less than $m + d$?

(A) 16%　　　(B) 32%　　　(C) 48%　　　(D) 84%　　　(E) 92%

解：

这是一个非常典型的标准正态分布。在一个标准差内是 68%，这说明在一个标准差之外还剩下 32%，由于标准正态分布的对称性，所以在 $m+d$ 外是 16%。题目问的是小于 $m+d$ 的占比，则为 100% − 16% = 84%。答案为 D。

例题 6

Of the numbers q, r, s and t, which is greatest?

(1) The average (arithmetic mean) of q and r is s.

(2) The sum of q and r is t.

(A) Statement (1) ALONE is sufficient, but statement (2) alone is not sufficient.

(B) Statement (2) ALONE is sufficient, but statement (1) alone is not sufficient.

(C) BOTH statements TOGETHER are sufficient, but NEITHER statement ALONE is sufficient.

(D) EACH statement ALONE is sufficient.

(E) Statements (1) and (2) TOGETHER are NOT sufficient.

解：

条件 1 说，q 和 r 的平均数是 s，从中我们无法知道 t 的大小，也无法知道 q 和 r 谁大，故条件 1 不充分。

条件 2 说，q 和 r 的和是 t。从中我们只能知道 t 比 q 和 r 大，但不知道 t 和 s 的情况，故条件 2 不充分。

两个条件同时成立时，无论 q 和 r 谁大，它们都比 t 小，所以这四个数中最大的一定是 t。故条件 1 + 条件 2 充分。

综上，答案为 C。

例题 7

If p, r and s are consecutive integers in ascending order and x is the average (arithmetic mean) of the three integers, what is the value of x?

(1) Twice x is equal to the sum of p, r, and s.

(2) The sum of p, r, and s is zero.

(A) Statement (1) ALONE is sufficient, but statement (2) alone is not sufficient.

(B) Statement (2) ALONE is sufficient, but statement (1) alone is not sufficient.

(C) BOTH statements TOGETHER are sufficient, but NEITHER statement ALONE is sufficient.

(D) EACH statement ALONE is sufficient.

(E) Statements (1) and (2) TOGETHER are NOT sufficient.

解:

题目问的是 3 个连续整数的平均数（$p < r < s$）。我们需要知道这三个数的和。

条件 1 说，$2x = p + r + s$。由平均数定义可知，$3x = p + r + s$。显然，此时 $3x = 2x$。当且仅当 $x = 0$ 时，能满足这个条件。因此，条件 1 充分。

条件 2 说，$p + r + s = 0$，显然条件 2 也是充分的。

综上，答案为 D。

例题 8

What is the median of the data set S that consists of the integers 17, 29, 10, 26, 15, and x?

(1) The average (arithmetic mean) of S is 17.

(2) The range of S is 24.

(A) Statement (1) ALONE is sufficient, but statement (2) alone is not sufficient.

(B) Statement (2) ALONE is sufficient, but statement (1) alone is not sufficient.

(C) BOTH statements TOGETHER are sufficient, but NEITHER statement ALONE is sufficient.

(D) EACH statement ALONE is sufficient.

(E) Statements (1) and (2) TOGETHER are NOT sufficient.

解:

题目问的是这些数的中位数，我们至少需要知道 x 和这些数的关系。

条件 1 说，S 的平均数是 17。显然，知道平均数可以求出 x 的值，自然可以知道这组数的中位数，故条件 1 充分。

条件 2 说，S 的极差是 24。现有数字的极差是 $29-10=19$。这表明，x 要么是最大的，要么是最小的。如果 x 是最大的数字，则 $x=34$，此时中位数为 $\dfrac{17+26}{2}$；如果 x 是最小的数字，则 $x=5$，此时中位数为 $\dfrac{15+17}{2}$。由于两种情况下中位数不相等，所以条件 2 不充分。

综上，答案为 A。

例题 9

X，81，73，71，98，73，64

What is the value of X in the above list of 7 numbers?

(1) The average (arithmetic mean) of these 7 numbers is 80.

(2) The range of these 7 numbers is 36.

(A) Statement (1) ALONE is sufficient, but statement (2) alone is not sufficient.

(B) Statement (2) ALONE is sufficient, but statement (1) alone is not sufficient.

(C) BOTH statements TOGETHER are sufficient, but NEITHER statement ALONE is sufficient.

(D) EACH statement ALONE is sufficient.

(E) Statements (1) and (2) TOGETHER are NOT sufficient.

解:

本题和上一道题是非常相似的。只不过要看清题目，题目问的是 X 的值。

条件 1 可以计算出 X 的值，故条件 1 充分。

通过条件 2 可求出 $X = 100$ 或 62，故条件 2 不充分。请注意，如果本题问的和上一道题一样，也是中位数的话，那么无论 X 是 100 还是 62，中位数均为 73，是充分的。

综上，答案为 A。

例题 10

Month	Average Price per Dozen
April	$1.26
May	$1.20
June	$1.08

The table above shows the average（arithmetic mean）price per dozen eggs sold in a certain store during three successive months. If $\dfrac{2}{3}$ as many dozen were sold in April as in May, and twice as many were sold in June as in April, what was the average price per dozen of the eggs sold over the three-month period?

（A）$1.08　　（B）$1.10　　（C）$1.14　　（D）$1.16　　（E）$1.18

解：

依题意，4 月:5 月:6 月 = 2:3:4。由此可以计算出总的平均价格（相当于加权平均），即

$$\frac{2 \times 1.26 + 3 \times 1.2 + 4 \times 1.08}{9} = 1.16。$$

故答案为 D。

例题 11

In a certain group of people, the average（arithmetic mean）weight of the males is 180 pounds and of the females is 120 pounds. What is the average weight of the people in the group?

（1）The group contains twice as many females as males.

（2）The group contains 10 more females than males.

(A) Statement (1) ALONE is sufficient, but statement (2) alone is not sufficient.

(B) Statement (2) ALONE is sufficient, but statement (1) alone is not sufficient.

(C) BOTH statements TOGETHER are sufficient, but NEITHER statement ALONE is sufficient.

(D) EACH statement ALONE is sufficient.

(E) Statements (1) and (2) TOGETHER are NOT sufficient.

解：

题目问的是所有人的平均体重，我们至少需要知道男性和女性的人数比值。

条件 1 说，女性是男性的 2 倍。显然，本选项给定了女性和男性的权重，总平均数应为：$\dfrac{180 + 2 \times 120}{2}$。故条件 1 充分。

条件 2 说，女性比男性多 10 人。本条件没有给出男性和女性的人数比，故条件 2 不充分。

综上，答案为 A。

例题 12

Each of the five divisions of a certain company sent representatives to a conference. If the numbers of representatives sent by four of the divisions were 3, 4, 5 and 5, was the range of the numbers of representatives sent by the five divisions greater than 2?

(1) The median of the numbers of representatives sent by the five divisions was greater than the average (arithmetic mean) of these numbers.

(2) The median of the numbers of representatives sent by the five divisions was 4.

(A) Statement (1) ALONE is sufficient, but statement (2) alone is not sufficient.

(B) Statement (2) ALONE is sufficient, but statement (1) alone is not sufficient.

(C) BOTH statements TOGETHER are sufficient, but NEITHER statement ALONE is sufficient.

(D) EACH statement ALONE is sufficient.

(E) Statements (1) and (2) TOGETHER are NOT sufficient.

解：

题目问的是极差是否大于 2。我们需要知道最后一个部门派出多少名代表，设其为 x。

条件 1 说 5 个数的中位数大于平均数。可以分两种情况讨论 x 取值。

情况 1：$x \leqslant 4$，此时，中位数为 4。即 $4 > \dfrac{17+x}{5}$，解得 $x < 3$。

情况 2：$x \geqslant 5$，此时，中位数是 5，即 $5 > \dfrac{17+x}{5}$，解得 $5 \leqslant x < 8$。

若 $x < 3$，极差大于 2；若 $5 \leqslant x < 8$，不能确定极差是否大于 2。

综上，条件 1 不充分。

条件 2 说中位数为 4。从中可以推出 $x \leqslant 4$，显然无法确定极差是否大于 2，故条件 2 不充分。

两个条件结合，交集为 $x < 3$，此时可以确定极差一定大于 2，故条件 1 + 条件 2 充分。

综上，答案为 C。

例题 13

In a numerical table with 10 rows and 10 columns, each entry is either a 9 or a 10. If the number of 9s in the n throw is $n-1$ for each n from 1 to 10, what is the average (arithmetic mean) of all the numbers in the table?

(A) 9.45   (B) 9.50   (C) 9.55   (D) 9.65   (E) 9.70

解：

本题的难度主要在于充分理解题意。题目的意思是：

有一个 10×10 的数字表格，每个格里要么是数字 9，要么是数字 10。如果在第 n 行，数字 9 的个数是 $n-1$（n 是从 $1 \sim 10$ 的任意一个数），那么这个数字表格的平均数（算数平均数）是多少？

第一行显然是 0 个 9，第二行是 1 个 9，第三行是 2 个 9 等，以此类推。

因此，算术平均数为：

$$\frac{100+99+98+97+96+95+94+93+92+91}{100}=9.55。$$

答案为 C。

例题 14

A certain list consists of 21 different numbers. If n is in the list and n is 4 times the average（arithmetic mean）of the other 20 numbers in the list，then n is what fraction of the sum of the 21 numbers in the list?

（A）$\frac{1}{20}$ （B）$\frac{1}{6}$ （C）$\frac{1}{5}$ （D）$\frac{4}{21}$ （E）$\frac{5}{21}$

解：

依题意，设其他 20 个数的算术平均数为 x，则 $n=4x$ 且其余 20 个数的和是 $20x$。

则 n 占 21 个数和的百分比为：

$$\frac{4x}{24x}=\frac{1}{6}。$$

答案为 B。

例题 15

A teacher gave the same test to three history classes：A，B and C. The average（arithmetic mean）scores for the three classes were 65，80 and 77，respectively. The ratio of the numbers of students in each class who took the test was 4 to 6 to 5，respectively. What was the average score for the three classes combined?

解：

这道题相当于一个加权平均的考题。65，80，77 的权重分别为 4，6，5，加权平均为：

$$\frac{65 \times 4 + 80 \times 6 + 77 \times 5}{15} = 75。$$

综上，答案为 B。

例题 16

If the average（arithmetic mean）of x, y and 20 is 10 greater than the average of x, y, 20 and 30，what is the average of x and y?

（A）40　　　（B）45　　　（C）60　　　（D）75　　　（E）95

解：

先算 x, y 和 20 的平均数，即

$$\frac{x + y + 20}{3},$$

再计算 x, y, 20 和 30 的平均数，即

$$\frac{x + y + 20 + 30}{4}。$$

依题意，则有：

$$\frac{x + y + 20}{3} - 10 = \frac{x + y + 20 + 30}{4}。$$

由此可知（解方程时，把 $x + y$ 当成一个整体来解），

$$x + y = 190。$$

两者的平均数为 95。综上，答案为 E。

例题 17

A certain list consists of 400 different numbers. Is the average（arithmetic mean）of the numbers in the list greater than the median of the numbers in the list?

（1）Of the numbers in the list, 280 are less than the average.

（2）Of the numbers in the list, 30 percent are greater than or equal to the average.

（A） Statement（1）ALONE is sufficient, but statement（2）alone is not sufficient.

（B） Statement（2）ALONE is sufficient, but statement（1）alone is not sufficient.

（C） BOTH statements TOGETHER are sufficient, but NEITHER statement ALONE is sufficient.

（D） EACH statement ALONE is sufficient.

（E） Statements（1）and（2）TOGETHER are NOT sufficient.

解：

条件 1 说，在 400 个数字中，有 280 个数比平均数小。根据中位数的定义，它肯定出现在这 280 个数中。由此可知，这列数的平均数大于中位数。故条件 1 充分。

条件 2 说，在 400 个数中，30% 的数是大于或等于平均数的。也就是说，70% 的数是小于平均数的。中位数的定义是从大到小排列后位于中间的数，所以中位数必然处于 70% 的数之中。也就是说，这列数的平均数大于中位数。故条件 2 充分。

综上，答案为 D。

例题 18

A set of 15 different integers has a median of 25 and a range of 25. What is the greatest possible integer that could be in this set?

（A）32   （B）37   （C）40   （D）43   （E）50

解：

中位数是 25，表示有 7 个数要小于等于 25。由于问的是最大的数且极差是确定的，所以这 7 个数应该尽量大。由于集合中所有数字均不相同，所以这 7 个数只能为 24，23，22，21，20，19，18。由于极差是 25，所以集合中最大的数字应为 $18+25=43$。

答案为 D。

2.9 ▸ 集合

2.9.1 ▸ 集合性质

所谓集合，就是由一个或多个确定的元素所构成的整体。例如，自然数集合为：

$$\{1,\ 2,\ 3,\ 4\cdots\}$$

集合有三个特性，分别为确定性、互异性和无序性。

确定性

给定一个集合，任意给出一个元素，该元素或者属于或者不属于该集合，二者必居其一，不允许有模棱两可的情况出现。

互异性

一个集合中，任何两个元素都认为是不相同的，即每个元素只能出现一次。有时需要对同一元素出现多次的情形进行刻画，可以使用多重集，其中的元素允许出现多次。

无序性

一个集合中，每个元素的地位都是相同的，元素之间是无序的。集合上可以定义序关系，定义了序关系后，元素之间就可以按照序关系排序。但就集合本身的特性而言，元素之间没有必然的序。

2.9.2 ▸ 子集和真子集

如果集合 A 的任意一个元素都是集合 B 的元素（任意 $a \in A$，则 $a \in B$），那么集合 A 称为集合 B 的子集，记为 $A \subseteq B$ 或 $B \supseteq A$，读作"集合 A 包含于集合 B"或"集合 B 包含集合 A"。

即：$\forall a \in A$ 有 $a \in B$，则 $A \subseteq B$。

如果集合 A 是 B 的子集，且 $A \neq B$，即 B 中至少有一个元素不属于 A，那么 A 就是 B 的真子集，可记作：$A \subsetneqq B$。

如果集合中一个元素都没有，我们称之为"空集"。对于空集 $\varnothing$，我们规定 $\varnothing \subseteq A$，即空集是任何集合的子集。

计算一个集合的子集个数和计算一个数的因数数量是同理的。对于有 n 个元素的集合来说，每一个元素均有出现或不出现在它的子集中这两种可能。因此，有 n 个元素的集合的子集个数为 2^n。

例题 1

A set of numbers has the property that for any number t in the set, $t + 2$ is in the set. If -1 is in the set, which of the following must also be in the set?

Ⅰ. -3　　　　Ⅱ. 1　　　　Ⅲ. 5

（A） Ⅰ only　　　（B） Ⅱ only　　　（C） Ⅰ and Ⅱ only

（D） Ⅱ and Ⅲ only　　（E） Ⅰ，Ⅱ and Ⅲ

解：

题目说，如果 t 在集合中，那么 $t + 2$ 就在集合中。但题目并没有说，$t + 2$ 在集合中，t 一定在集合中。

因此，如果 -1 在，那么 1 在；1 在，则 3 在；3 在，则 5 在。至于 -1 是怎么来的，题目没说，不能自行推理。

综上，答案为 D。

例题 2

The subsets of the set $\{w, x, y\}$ are $\{w\}$, $\{x\}$, $\{y\}$, $\{w, x\}$, $\{w, y\}$, $\{x, y\}$, $\{w, x, y\}$, and $\{\ \}$ (the empty subset). How many subsets of the set $\{w, x, y, z\}$ contain w?

（A）four （B）five （C）seven （D）Eight （E）sixteen

解：

这道题有点像之前讲过的因数数量问题。我们可以这么想，对于包含 w 的集合（子集）来说，x 是可能有，可能没有的，共两种可能；y 和 z 同理。因此，包含 w 的子集的情况为：

$$2 \times 2 \times 2 = 8。$$

故答案为 D。

例题 3

The cardinality of a finite set is the number of elements in the set. What is the cardinality of set A?

（1） 2 is the cardinality of exactly 6 subsets of set A.

（2） Set A has a total of 16 subsets, including the empty set and set A itself.

（A） Statement （1） ALONE is sufficient, but statement （2） alone is not sufficient.

（B） Statement （2） ALONE is sufficient, but statement （1） alone is not sufficient.

（C） BOTH statements TOGETHER are sufficient, but NEITHER statement ALONE is sufficient.

（D） EACH statement ALONE is sufficient.

（E） Statements （1） and （2） TOGETHER are NOT sufficient.

解：

题目问的是集合 A 中含有的元素数量。

条件 1 说，集合 A 中恰好有两个元素的子集是 6 个。这里需要借助一下排列组合的知识。集合 A 的有两个元素的子集，其实就是从集合 A 的 n 个元素中选取 2 个元素的所有可能性，即

$$C_n^2 = 6,$$

解出 $n = 4$。故条件 1 充分。

条件 2 说，集合 A 有总共 16 个子集。而含有 n 个元素的有限集合的子集个数为 2^n。

由此可知，$2^n = 16$，即 $n = 4$。故条件 2 充分。

综上，答案为 D。

例题 4

If S is a set of odd integers and 3 and -1 are in S, is -15 in S?

（1）5 is in S.

（2）Whenever two numbers are in S, their product is in S.

（A）Statement（1）ALONE is sufficient, but statement（2）alone is not sufficient.

（B）Statement（2）ALONE is sufficient, but statement（1）alone is not sufficient.

（C）BOTH statements TOGETHER are sufficient, but NEITHER statement ALONE is sufficient.

（D）EACH statement ALONE is sufficient.

（E）Statements（1）and（2）TOGETHER are NOT sufficient.

解：

条件 1 说，5 在集合 S 中，由此我们无法判断 -15 的情况，故条件 1 不充分。

条件 2 说，如果两个数在 S 中，那么他们的积就在 S 中。题干中给出的集合元素只有 3 和 -1，没有 5，所以无论怎么乘，也不能判断 -15 的情况，故条件 2 不充分。

两个条件同时成立时，显然，可以确认 -15 一定在集合 S 中，故条件 1 + 条件 2 充分。

综上，答案为 C。

2.9.3 ▶ 容斥原理

讲到集合，一定会想到韦恩图（文氏图）。讲到韦恩图，最常见的考点必是容斥原理。

两个集合的容斥原理比较简单:

$A \cup B = |A \cup B| = |A| + |B| - |A \cap B|$ ($\cap$: 相交重合的部分)

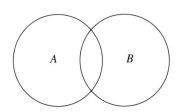

三个集合的容斥原理相对复杂些:

$$|A \cup B \cup C| = |A| + |B| + |C| - |A \cap B| - |B \cap C| - |C \cap A| + |A \cap B \cap C|$$

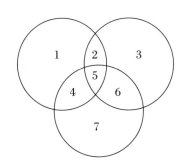

三个集合的容斥原理相对较难以理解,此处给出推理过程:

$$等式右边 = \{ [(A + B - A \cap B) + C - B \cap C] - C \cap A \} + A \cap B \cap C。$$

韦恩图分块标记如上图所示: 1245 构成 A, 2356 构成 B, 4567 构成 C。

等式右边( )里指的是上图中的 1 + 2 + 3 + 4 + 5 + 6 六部分: 那么 $A \cup B \cup C$ 还缺部分 7。

等式右边[ ]号里 + C (即 4 + 5 + 6 + 7 四个部分) 后, 相当于 $A \cup B \cup C$ 多加了 4 + 5 + 6 三部分, 减去 $B \cap C$ (即 5 + 6 两部分) 后, 还多加了部分 4。

等式右边{ }里减去 $C \cap A$ (即 4 + 5 两部分) 后, $A \cup B \cup C$ 又多减了部分 5, 则加上 $A \cap B \cap C$ (即部分 5) 刚好是 $A \cup B \cup C$。

例题 5

70 percent of workers used computers to surf the Internet. 80 percent of the workers used computers to read novels. 10 percent of those workers neither used computers to surf the Internet nor used computers to read novels. What percent of those workers used computers both to surf the Internet and to read novels?

（A）80%　　　（B）70%　　　（C）60%　　　（D）50%　　　（E）40%

解：

因为有10%的人既不上网又不读小说，所以一共用电脑做事的人一共是90%。基于容斥原理，$A \cup B = |A \cup B| = |A| + |B| - |A \cap B|$，即

$$90\% = 70\% + 80\% - A \cap B,$$

$A \cap B = 60\%$，故答案为 C。

例题 6

Last year 26 members of a certain club traveled to England, 26 members traveled to France, and 32 members traveled to Italy. Last year no members of the club traveled to both England and France, 6 members traveled to both England and Italy, and 11 members traveled to both France and Italy. How many members of the club traveled to at least one of these three countries last year?

（A）52　　　（B）67　　　（C）71　　　（D）73　　　（E）79

解：

依题意得：

$$E = 26；F = 26；I = 32；$$

$$E \cap F = 0；E \cap I = 6；F \cap I = 11；E \cap F \cap I = 0。$$

根据容斥原理公式：

$E \cup F \cup I = 26 + 26 + 32 - 0 - 6 - 11 + 0 = 67$，故答案为 B。

定量推理简介
第一章

算数
第二章

代数
第三章

几何
第四章

文字问题
第五章

例题 7

All trainees in a certain aviator training program must take both a written test and a flight test. If 70 percent of the trainees passed the written test, and 80 percent of the trainees passed the flight test, what percent of the trainees passed both tests?

(1) 10 percent of the trainees did not pass either test.

(2) 20 percent of the trainees passed only the flight test.

(A) Statement (1) ALONE is sufficient, but statement (2) alone is not sufficient.

(B) Statement (2) ALONE is sufficient, but statement (1) alone is not sufficient.

(C) BOTH statements TOGETHER are sufficient, but NEITHER statement ALONE is sufficient.

(D) EACH statement ALONE is sufficient.

(E) Statements (1) and (2) TOGETHER are NOT sufficient.

解：

条件 1 说，10% 的人两项测试都没通过。这个条件告诉我们，通过某一项测试或两项测试都通过的人为 90%。根据容斥原理公式：

$$A \cup B = |A \cup B| = |A| + |B| - |A \cap B|,$$

则有：

$$90\% = 70\% + 80\% - |A \cap B|,$$

由此可得到 $|A \cap B|$ 的值，故条件 1 充分。

条件 2 说，20% 的人只通过了飞行测试。因为有 80% 的人通过了飞行测试（包含两项测试都通过的），所以两项测试都通过的人为：

$$80\% - 20\% = 60\%。$$

故条件 2 充分。

综上，答案为 D。

Of the 500 business people surveyed, 78 percent said that they use their laptop computers at home, 65 percent said that they use them in hotels, and 52 percent said that they use them both at home and in hotels. How many of the business people surveyed said that they do not use their laptop computers either at home or in hotels?

（A）45　　　（B）55　　　（C）65　　　（D）95　　　（E）130

解：

容斥原理公式：

$$A \cup B = |A \cup B| = |A| + |B| - |A \cap B|。$$

设在家里用笔记本电脑的人为 A，在酒店里用笔记本电脑的人为 B，则有：

$$|A \cup B| = |78\% \times 500| + |65\% \times 500| - |52\% \times 500| = 455。$$

题目问的是在两个地方都不用笔记本电脑的人，则有：

$$500 - 455 = 45。$$

答案为 A。

What is the number of integers that are common to both set S and set T?

（1）The number of integers in S is 7, and the number of integers in T is 6.

（2）U is the set of integers that are in S only or in T only or in both, and the number of integers in U is 10.

（A）Statement（1）ALONE is sufficient, but statement（2）alone is not sufficient.

（B）Statement（2）ALONE is sufficient, but statement（1）alone is not sufficient.

（C）BOTH statements TOGETHER are sufficient, but NEITHER statement ALONE is sufficient.

（D）EACH statement ALONE is sufficient.

（E）Statements（1）and（2）TOGETHER are NOT sufficient.

解:

题目问的是两个集合的交集。

条件 1 给出了两个集合各自的情况,无法判断交集的情况,故条件 1 不充分。

条件 2 说 U 是仅在集合 S 或仅在集合 T 或仅在 S 和 T 交集里的一组数,U 的整数数量为 10。仅凭此条件,无法判断 U 是否是交集,故条件 2 不充分。

当两个条件同时成立时,因为 U 里的整数的数量比 S 和 T 都多,所以 U 只能是交集。故条件 1 + 条件 2 充分。

综上,答案为 C。

例题 10

Three thousand families live in a certain town. How many families who live in the town own neither a car nor a television set?

(1) Of the families who live in the town, 2,980 own a car.

(2) Of the families who live in the town, 2,970 own both a car and a television set.

(A) Statement (1) ALONE is sufficient, but statement (2) alone is not sufficient.

(B) Statement (2) ALONE is sufficient, but statement (1) alone is not sufficient.

(C) BOTH statements TOGETHER are sufficient, but NEITHER statement ALONE is sufficient.

(D) EACH statement ALONE is sufficient.

(E) Statements (1) and (2) TOGETHER are NOT sufficient.

解:

条件 1 说,住在镇上的家庭有 2980 个有小汽车。其中没有提到电视机的情况,故条件 1 不充分。

条件 2 说,住在镇上的家庭有 2970 个既有小汽车,又有电视机。因为不知道只有电视机和只有汽车的家庭数量,所以无法得知两者都没有的情况。故条件 2 不充分。

两个条件同时成立时，我们不知道只有电视机的家庭的数量，所以也无法求得两者都没有的情况。

综上，答案为 E。

例题 11

Sets A, B, and C have some elements in common. If 16 elements are in both A and B, 17 elements are in both A and C, and 18 elements are in both B and C, how many elements do all three of the sets A, B, and C have in common?

(1) Of the 16 elements that are in both A and B, 9 elements are also in C.

(2) A has 25 elements, B has 30 elements, and C has 35 elements.

(A) Statement (1) ALONE is sufficient, but statement (2) alone is not sufficient.

(B) Statement (2) ALONE is sufficient, but statement (1) alone is not sufficient.

(C) BOTH statements TOGETHER are sufficient, but NEITHER statement ALONE is sufficient.

(D) EACH statement ALONE is sufficient.

(E) Statements (1) and (2) TOGETHER are NOT sufficient.

解：

题目问的是 $|A \cap B \cap C|$。

条件 1 说，在 A 和 B 共有的 16 个元素中，有 9 个也在 C 里。这就表明 $|A \cap B \cap C|$ =9，故条件 1 充分。

条件 2 告诉了我们集合 A，B 和 C 各自的元素数，根据容斥原理公式，即

$$|A \cup B \cup C| = |A| + |B| + |C| - |A \cap B| - |B \cap C| - |C \cap A| + |A \cap B \cap C|,$$

可以发现，本条件中没有给出 $|A \cup B \cup C|$ 的值，因此无法求解。故条件 2 不充分。

综上，答案为 A。

例题 12

How many people in Town X read neither the *World* newspaper nor the *Globe* newspaper?

(1) Of the 2,500 people in Town X, 1,000 read no newspaper.

(2) Of the people in Town X, 700 read the *Globe* only and 600 read the *World* only.

(A) Statement (1) ALONE is sufficient, but statement (2) alone is not sufficient.

(B) Statement (2) ALONE is sufficient, but statement (1) alone is not sufficient.

(C) BOTH statements TOGETHER are sufficient, but NEITHER statement ALONE is sufficient.

(D) EACH statement ALONE is sufficient.

(E) Statements (1) and (2) TOGETHER are NOT sufficient.

解：

条件 1 说，在 X 镇的 2500 个人中，1000 个人不读报纸。我们不知道这个镇是否还有其他报纸，所以不能认为这 1000 个人就是题干中提到的两种报纸都不读的人数，即

{不读报纸的人} = {不读 *World*、不读 *Globe* 且不读其他报纸的人} ⊆ {不读 *World* 且不读 *Globe* 的人}，故条件 1 不充分。

条件 2 说，所有人中，700 个人只读 *Globe*，600 个人只读 *World*。这个条件没告诉我们总共的人数和交集的人数，故条件 2 不充分。

两个条件同时成立时，显然依然无法确定两种报纸都不读的人数（依然是因为不知道是否有其他报纸）。

综上，答案为 E。

Results of a Poll

Company	Number of Those Who Own Stock in the Company
AT&/T	30
IBM	48
GM	54
FORD	75
US Air	83

In a poll, 200 subscribers to *Financial Magazine* X indicated which of five specific companies they own stock in. The results are shown in the table above. If 15 of the 200 own stock in both IBM and AT&T, how many of those polled own stock in neither company?

(A) 63 (B) 93 (C) 107 (D) 122 (E) 137

解:

依题意，利用容斥原理的公式则有:

$$A \cup B = |A \cup B| = |A| + |B| - |A \cap B|,$$

$$AT\&T \cup IBM = 30 + 48 - 15 = 63_\circ$$

两个公司都没有股票的人数为 $200 - 63 = 137_\circ$

答案为 E。

What is the number of integers that are common to both set *S* and set *T*?

(1) The number of integers in *S* is 7, and the number of integers in *T* is 6.

(2) *U* is the set of integers that are in *S* only or in *T* only or in both, and the number of integers in *U* is 10.

(A) Statement (1) ALONE is sufficient, but statement (2) alone is not sufficient.

(B) Statement (2) ALONE is sufficient, but statement (1) alone is not sufficient.

（C）BOTH statements TOGETHER are sufficient, but NEITHER statement ALONE is sufficient.

（D）EACH statement ALONE is sufficient.

（E）Statements（1）and（2）TOGETHER are NOT sufficient.

解：

题目问的是两个集合的交集。

条件 1 给出了两个集合中各自整数的数量，显然无法求得交集，故条件 1 不充分。

条件 2 给出的 U 是集合 S 和集合 T 的并集，显然无法求得交集，故条件 2 不充分。

两个条件同时成立时，根据容斥原理公式则有：

$$U = S + T - S \cap T = 10,$$

解得 $S \cap T = 3$，故条件 1 + 条件 2 充分。

综上，答案 C。

2.9.4 ▶ 集合元素有两种属性

如果集合中所有元素都只有一个属性，那么用韦恩图来表达是完美的。但如果集合中的元素都有不止一个属性，韦恩图就会因为只能表达一个维度而完全失效。当题目涉及的集合元素有两个维度时，我们就必须用"列表法"来解题。

例题 15

In a certain production lot, 40 percent of the toys are red and the remaining toys are green. Half of the toys are small and half are large. If 10 percent of the toys are red and small, and 40 toys are green and large, how many of the toys are red and large?

（A）20   （B）30   （C）40   （D）50   （E）60

解：

本题中每个玩具都有两个属性，即大小和颜色。为了方便解题，我们用一个表格来表示整个集合的情况，如下：

	Red	Green	Total
Small	10%		50%
Large			50%
Total	40%	60%	100%

通过简单的加减运算，可将表填满，则有：

	Red	Green	Total
Small	10%	40%	50%
Large	30%	20%	50%
Total	40%	60%	100%

因为 20% 的玩具是又大又绿的，因此 $20\% \times$ 总数 $=40$；由此可知，总数为 200。又大又红的玩具占 30%，因此，它们的总数为 $30\% \times 200 = 60$，答案为 E。

2.10 ▸ 比大小问题

顾名思义，比大小问题是选项中给出不同的数字，要求考生比较这些数字的大小。比大小考题的核心在于"设立统一标准"。

例如，如果想直接比较 2^4 和 9^2 是相对困难的，但我们可以让它们的幂指数均变为 4，即 $9^2 = 3^4$。同幂指数下，比较大小就变得非常直观了。在这个例子中，"幂指数相同"就是"统一的标准"。

例题 1

Which of the following fractions has the greatest value?

(A) $\dfrac{1}{3^2 5^2}$ (B) $\dfrac{2}{3^2 5^2}$ (C) $\dfrac{7}{3^3 5^2}$ (D) $\dfrac{45}{3^3 5^3}$ (E) $\dfrac{75}{3^4 5^5}$

解：

我们可以将分母统一为最大的数字，即 $3^4 5^5$。接下来只需比较分子即可，分子越大，数字越大。

选项 A 的分子为：$3^2 \times 5^3$，

选项 B 的分子为：$2 \times 3^2 \times 5^3$，

选项 C 的分子为：$7 \times 3 \times 5^3$，

选项 D 的分子为：$45 \times 3 \times 5^2$，

选项 E 的分子为：75。

显然，选项 D 的分子最大。答案为 D。

例题 2

If $x<0$ and $0<y<1$, which of the following has the greatest value?

(A) x^2 (B) $(xy)^2$ (C) $\left(\dfrac{x}{y}\right)^2$ (D) $\dfrac{x^2}{y}$ (E) $x^2 y$

解：

观察题目可知，我们可以首先以 x^2 为标准，由于 $0<y<1$，所以 x^2 乘 y，肯定是越乘越小；除以 y，肯定是越除越大。由此可知，

$$\frac{x^2}{y^2} > \frac{x^2}{y} > x^2 > x^2 y > x^2 y^2。$$

综上，答案为 C。

例题 3

If $-1<h<0$, which of the following has the greatest value?

(A) $1-h$ (B) $1+h$ (C) $1+h^2$ (D) $1-\dfrac{1}{h}$ (E) $1-\dfrac{1}{h^2}$

解：

先以 1 为标准。因为 $-1 < h < 0$，所以 $1 - h > 1$；$1 + h > 1$；$1 + h^2 > 1$；$1 - \dfrac{1}{h} > 1$；

$1 - \dfrac{1}{h^2} < 1$。

此时可以先排除 E 选项。

再以 $1 + h$ 为标准，因为 h 是负数，所以 $1 - h > 1 + h$；$1 + h^2 > 1 + h$；$1 - \dfrac{1}{h} > 1 + h$。

最后以 $1 - \dfrac{1}{h}$ 为标准，因为 $-1 < h < 0$，所以 $1 - \dfrac{1}{h} > 1 - h$；$1 - \dfrac{1}{h} > 1 + h^2$。

综上，答案为 D。

例题 4

If $0 < x < 1$, what is the median of the values x, x^{-1}, x^2, $\sqrt{x}$ and x^3?

(A) x (B) x^{-1} (C) x^2 (D) $\sqrt{x}$ (E) x^3

解：

这道题问的是中位数，其实也是个比大小的问题。以 x 为标准，因为 $0 < x < 1$，所以 $x > x^2 > x^3$；同时，$x^{-1} > x$ 且 $\sqrt{x} > x$。由此可知，x 必然为这五个数中最中间的数。

因此，答案为 A。

1. If $S = \{0, 4, 5, 2, 11, 8\}$, how much greater than the median of the numbers in S is the mean of the numbers in S?

    (A) 0.5   (B) 1.0   (C) 1.5   (D) 2.0   (E) 2.5

2. If a and b are positive integers and $(2^a)^b = 23$, what is the value of $2^a 2^b$?

    (A) 6   (B) 8   (C) 16   (D) 32   (E) 64

3. $\dfrac{(-1.5)(1.2) - (4.5)(0.4)}{30} =$

    (A) -1.2 (B) -0.12 (C) 0   (D) 0.12   (E) 1.2

4. If the average (arithmetic mean) of x, y and z is $7x$ and $x \neq 0$, what is the ratio of x to the sum of y and z?

    (A) 1:21   (B) 1:20   (C) 1:6   (D) 6:1   (E) 20:1

5. Of all the students in a certain dormitory, $\dfrac{1}{2}$ are first-year students and the rest are second-year students. If $\dfrac{4}{5}$ of the first-year students have not declared a major and if the fraction of second-year students who have declared a major is 3 times the fraction of first-year students who have declared a major, what fraction of all the students in the dormitory are second-year students who have not declared a major?

    (A) $\dfrac{1}{15}$ (B) $\dfrac{1}{5}$ (C) $\dfrac{4}{15}$ (D) $\dfrac{1}{3}$ (E) $\dfrac{2}{5}$

6. If x and y are integers such that $2 < x \leq 8$ and $2 < y \leq 9$, what is the maximum value of $\dfrac{1}{x} - \dfrac{x}{y}$?

    (A) $-\dfrac{25}{8}$ (B) 0   (C) $\dfrac{1}{4}$ (D) $\dfrac{5}{18}$ (E) 2

7. Which of the following must be equal to zero for all real numbers x?

    I. $-\dfrac{1}{x}$ II. $x + (-x)$ III. x^0

(A) I only   (B) II only   (C) I and III only

(D) II and III only   (E) I, II, and III

8. If r and s are positive integers such that $(2^r)(4^s) = 16$, then $2r + s =$

(A) 2   (B) 3   (C) 4   (D) 5   (E) 6

9. In which of the following pairs are the two numbers reciprocals of each other?

$$\text{I. } 3 \text{ and } \frac{1}{3} \qquad \text{II. } \frac{1}{17} \text{ and } -\frac{1}{17} \qquad \text{III. } \sqrt{3} \text{ and } \frac{\sqrt{3}}{3}$$

(A) I only   (B) II only   (C) I and II

(D) I and III   (E) II and III

10. Of the following, which is least?

(A) $\dfrac{0.03}{0.00071}$ (B) $\dfrac{0.03}{0.0071}$ (C) $\dfrac{0.03}{0.071}$

(D) $\dfrac{0.03}{0.71}$ (E) $\dfrac{0.03}{7.1}$

11. Michael arranged all his books in a bookcase with 10 books on each shelf and no books left over. After Michael acquired 10 additional books, he arranged all his books in a new bookcase with 12 books on each shelf and no books left over. How many books did Michael have before he acquired the 10 additional books?

(1) Before Michael acquired the 10 additional books, he had fewer than 96 books.

(2) Before Michael acquired the 10 additional books, he had more than 24 books.

(A) Statement (1) ALONE is sufficient, but statement (2) alone is not sufficient.

(B) Statement (2) ALONE is sufficient, but statement (1) alone is not sufficient.

(C) BOTH statements TOGETHER are sufficient, but NEITHER statement ALONE is sufficient.

(D) EACH statement ALONE is sufficient.

(E) Statements (1) and (2) TOGETHER are NOT sufficient.

12. Is the positive integer n odd?

(1) $n^2 + (n+1)^2 + (n+2)^2$ is even.

(2) $n^2 - (n+1)^2 - (n+2)^2$ is even.

(A) Statement (1) ALONE is sufficient, but statement (2) alone is not sufficient.

(B) Statement (2) ALONE is sufficient, but statement (1) alone is not sufficient.

(C) BOTH statements TOGETHER are sufficient, but NEITHER statement ALONE is sufficient.

(D) EACH statement ALONE is sufficient.

(E) Statements (1) and (2) TOGETHER are NOT sufficient.

13. A small school has three foreign language classes, one in French, one in Spanish, and one in German. How many of the 34 students enrolled in the Spanish class are also enrolled in the French class?

(1) There are 27 students enrolled in the French class, and 49 students enrolled in either the French class, the Spanish class, or both of these classes.

(2) One-half of the students enrolled in the Spanish class are enrolled in more than one foreign language class.

(A) Statement (1) ALONE is sufficient, but statement (2) alone is not sufficient.

(B) Statement (2) ALONE is sufficient, but statement (1) alone is not sufficient.

(C) BOTH statements TOGETHER are sufficient, but NEITHER statement ALONE is sufficient.

(D) EACH statement ALONE is sufficient.

(E) Statements (1) and (2) TOGETHER are NOT sufficient.

14. If S is a set of four numbers w, x, y and z, is the range of the numbers in S greater than 2?

(1) $w - z > 2$

(2) z is the least number in S.

(A) Statement (1) ALONE is sufficient, but statement (2) alone is not sufficient.

(B) Statement (2) ALONE is sufficient, but statement (1) alone is not sufficient.

(C) BOTH statements TOGETHER are sufficient, but NEITHER statement ALONE is sufficient.

(D) EACH statement ALONE is sufficient.

(E) Statements (1) and (2) TOGETHER are NOT sufficient.

15. For the 5 days shown in the graph, how many kilowatt-hours greater was the median daily electricity use than the average (arithmetic mean) daily electricity use?

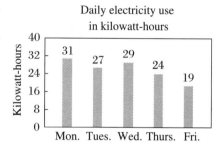

Daily electricity use
in kilowatt-hours

(A) 1   (B) 2   (C) 3

(D) 4   (E) 5

16. The product P of two prime numbers is between 9 and 55. If one of the prime numbers is greater than 2 but less than 6 and the other is greater than 13 but less than 25, then $P =$

(A) 15   (B) 33   (C) 34   (D) 46   (E) 51

17. What is the greatest common divisor of positive integers m and n ?

(1) m is a prime number.

(2) $2n = 7m$

(A) Statement (1) ALONE is sufficient, but statement (2) alone is not sufficient.

(B) Statement (2) ALONE is sufficient, but statement (1) alone is not sufficient.

(C) BOTH statements TOGETHER are sufficient, but NEITHER statement ALONE is sufficient.

(D) EACH statement ALONE is sufficient.

(E) Statements (1) and (2) TOGETHER are NOT sufficient.

18. If k is an integer greater than 1, is k equal to $2r$ for some positive integer r?

(1) k is divisible by 2^6.

(2) k is not divisible by any odd integer greater than 1.

(A) Statement (1) ALONE is sufficient, but statement (2) alone is not sufficient.

(B) Statement (2) ALONE is sufficient, but statement (1) alone is not sufficient.

(C) BOTH statements TOGETHER are sufficient, but NEITHER statement ALONE is sufficient.

(D) EACH statement ALONE is sufficient.

(E) Statements (1) and (2) TOGETHER are NOT sufficient.

19.■ If $\frac{2}{5}$ of the students at College C are business majors, what is the number of female students at College C?

(1) $\frac{2}{5}$ of the male students at College C are business majors.

(2) 200 of the female students at College C are business majors.

(A) Statement (1) ALONE is sufficient, but statement (2) alone is not sufficient.

(B) Statement (2) ALONE is sufficient, but statement (1) alone is not sufficient.

(C) BOTH statements TOGETHER are sufficient, but NEITHER statement ALONE is sufficient.

(D) EACH statement ALONE is sufficient.

(E) Statements (1) and (2) TOGETHER are NOT sufficient.

20.■ The arithmetic mean and standard deviation of a certain normal distribution are 13.5 and 1.5, respectively. What value is exactly 2 standard deviations less than the mean?

(A) 10.5　　(B) 11.0　　(C) 11.5　　(D) 12.0　　(E) 12.5

1. A。

首先把集合 S 按照大小顺序排列：$S = \{0, 2, 4, 5, 8, 11\}$，中位数 $= \dfrac{4+5}{2} = 4.5$；平均

数 $= \dfrac{0+2+4+5+8+11}{6} = 5$；中位数与平均数的差 $= 5 - 4.5 = 0.5$。

2. C。

$\left(2^a\right)^b = 2^{ab} = 2^3$，所以 $ab = 3$。因为 a，b 是整数，所以 $a = 1$，$b = 3$；或者 $a = 3$，$b = 1$。
这两种情况下：$a + b = 4$，$2^a \times 2^b = 2^{a+b} = 2^4 = 16$。

3. B。

略。

4. B。

$7x = \dfrac{x+y+z}{3}$，整理得：$y + z = 20x$。

5. B。

用表格可以清晰明了地看出各种情况的比例。假设总人数为 M，整理出下表：

	选专业	未选专业	总和
一年级学生数量	$\dfrac{1}{5} \times \dfrac{1}{2} \times M$	$\dfrac{4}{5} \times \dfrac{1}{2} \times M$	$\dfrac{1}{2} M$
二年级学生数量	$3 \times \dfrac{1}{5} \times \dfrac{1}{2} \times M$	$\dfrac{2}{5} \times \dfrac{1}{2} \times M$	$\dfrac{1}{2} M$
总和	$\dfrac{2}{5} \times M$	$\dfrac{3}{5} \times M$	M

二年级未选专业的学生数 $= \dfrac{2}{5} \times \dfrac{1}{2} \times M = \dfrac{1}{5} M$。

6. B。

$\dfrac{1}{x} - \dfrac{x}{y}$ 最大，需要满足 $\dfrac{1}{x}$ 最大且 $\dfrac{x}{y}$ 最小，$x = 3$ 时，$\dfrac{1}{x}$ 最大。$\dfrac{x}{y}$ 最小，应满足：$x = 3$，

$y = 9$（分子最小，分母最大），$x = 3$ 可以同时满足 $\dfrac{1}{x}$ 最大，$\dfrac{x}{y}$ 最小。所以，$x = 3$，

$y = 9$ 时，$\dfrac{1}{x} - \dfrac{x}{y}$ 是最大值，最大值 $= \dfrac{1}{3} - \dfrac{3}{9} = 0$。

7. B。

$-\dfrac{1}{x}$ 是分数，所以肯定不是 0；$x + (-x) = x - x = 0$；$x^0 = 1$，只有 Ⅱ 是 0。

8. D。

$(2^r)(4^s) = 2^r \times 2^{2s} = 2^{r+2s} = 16 = 2^4$。因此，$r + 2s = 4$。因为 r 与 s 都是正整数，$r = 4 - 2s > 0$，$s < 2$，s 只能等于 1；$r = 2$。综上 $2r + s = 4 + 1 = 5$。

9. D。

略。

10. E。

分子相同的两个分数，分母小的分数比较大。反过来说，分子相同的两个分数，分母大的分数比较小。

11. A。

设 Michael 在获得 10 本额外的书之前有 x 本书，根据题干描述，需要 x 能被 10 整除，$x + 10$ 能被 12 整除。

条件 1 说，在获得 10 本额外书之前的书数量小于 96，即 $x < 96$。满足题干要求中能被 10 整除的只有 10，20，30，40，50，60，70，80，90 共 9 种情况，又要满足 $x + 10$ 能被 12 整除，只有 $x = 50$ 满足要求，题目得解，故条件 1 充分。

条件 2 说，x 是能被 10 整除的数，又因为满足 $x > 24$，x 可以是 30，40，50，…，甚至是 500，…，所以无法确定书的数量，故条件 2 不充分。

12. D。

判断 n 的奇偶性。

对于条件 1 采用假设法，如果 n 是奇数，看条件 1 中的 $n^2 + (n+1)^2 + (n+2)^2$ 是否是偶数。如果 n 是奇数，则 n^2 是奇数，$(n+1)$ 是偶数，$(n+1)^2$ 是偶数，$(n+2)$ 是奇数，$(n+2)^2$ 是奇数，奇数 + 偶数 + 奇数 = 偶数，因此，n 是奇数成立，条

定量推理简介 第一章

算数 第二章

代数 第三章

几何 第四章

文字问题 第五章

件 1 充分。

对于条件 2，$(n+1)^2$ 是偶数，$-(n+1)^2$ 也是偶数，$(n+2)^2$ 是奇数，$-(n+2)^2$ 也是奇数，同样，奇数 + 偶数 + 奇数 = 偶数，因此，n 是奇数成立，故条件 2 充分。

13⊫⊪ A。

下图中表示的集合为：左边椭圆表示参加西班牙语班的学生，右边表示参加法语班的学生，中间相交的部分表示既参加西班牙语班又参加法语班的学生，设其为 y，只参加西班牙语班的学生数为 x，只参加法语班的学生数为 z。已知西班牙语班的学生人数为 $x+y=34$。

条件 1 说，27 人参加了法语班，即 $y+z=27$，有 49 名学生参加了法语班、西班牙语班或两个班都参加了，即 $x+y+z=49$，因为 $x+y=34$，所以 $z=15$，$y=12$，即有 12 个人同时参加了西班牙语班和法语班。故条件 1 充分。

条件 2 说，参加西班牙语班的 34 人中，有 17 人参加了多语言课程，但不能确定具体参加法语班和德语班的有多少人，所以无法计算，故条件 2 不充分。

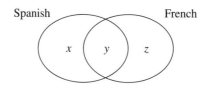

14⊫⊪ A。

条件 1 说，$w-z>2$，说明已经存在两个数的差值大于 2，其余数无论差值多少，$w-z>2$ 已经能保证集合 S 的范围大于 2，故条件 1 充分。

条件 2 说，只知道 z 是最小值，无法得知其他数值的大小，就无法计算范围，也不能确定范围是否大于 2。故条件 2 不充分。

15⊫⊪ A。

本题是考查中位数和算数平均数的题目。

先将 5 个数值按顺序排列：19，24，27，29，31，得到中位数为 27。

算数平均数为：$\dfrac{19+24+27+29+31}{5}=26$。

题目问：日用电量的中位数比平均数大多少，则 $27-26=1$。

定量推理简介
第一章

算数
第二章

代数
第三章

几何
第四章

文字问题
第五章

16 ━━ E。

大于 2 小于 6 的质数有 3 和 5，大于 13 小于 25，有 17，19，23，乘积最小为：$3 \times 17 = 51$，只有这个数落在了 $9 \sim 55$ 这个区间之内。

17 ━━ C。

条件 1 说，m 是质数。质数的约数只有 1 和它本身，但不知道 n 的情况，所以无法确定它们的最大公约数。故条件 1 不充分。

条件 2 说，$2n = 7m$，$2n$ 一定是偶数，$7m$ 也是偶数，所以 m 一定是一个偶数。但依然不能得出两者的最大公约数。故条件 2 不充分。

条件 1 + 条件 2：m 是质数，同时是偶数。我们知道，同时是质数还是偶数的数字只有 2，所以 $m = 2$，$n = 7$，最大公约数是 1，故条件 1 + 条件 2 充分。

18 ━━ B。

条件 1 说，k 能被 26 整除。举一个反例：$k = 3 \times 26$，k 此时不等于 $2r$。当 $k = 2r \times 2n$，n 整数时，原问题成立，所以条件 1 不充分。

条件 2 说，k 不能被任何大于 1 的奇数整除，这也就是说，k 的质因数都是偶数，是偶数的质数只有一个 2，所以 k 一定是 2 的整数次幂，所以条件 2 充分。

注意：数字分成两种：质数和合数。K 不能被任何一个奇数整除，所有除了 2 以外的质数都是奇数。如果 K 是质数，那么此时 K 只能是 2，否则 K 就多了一个奇数质因数——它本身。如果 K 是合数，那么 K 中含有的质因数只能是 2，所以，依然可以表达为 $2r$。

19 ━━ C。

条件 1 说，$\frac{2}{5}$ 的男生是学商科的。但并不知道具体人数是多少，也就无法求得女生的人数，故条件 1 不充分。

条件 2 说，200 个女生是学商科的，但并不知道商科中男生的人数，所以无法求得总学生数，也就不知道其余专业中男女生各是多少，故条件 2 不充分。

条件 1 + 条件 2：$\frac{2}{5}$ 的男生学商科，则 $\frac{3}{5}$ 的男生学其他的专业，学生总人数的 $\frac{2}{5}$ 学商科，那么学商科的女生就占到女生总人数的 $\frac{2}{5}$，又知学商科的女生有 200 人，所以女生总人数是 500 人。所以条件 1 + 条件 2 充分。

20 ◼◖◖ A。

平均值 13.5 减去 2 倍的标准差 1.5，即 13.5 - 3 = 10.5。

第三章

——

代 数

3.1 ▸ 因式分解与化简

因式分解只需记住如下四个规则或公式即可。

平方差公式:

$$a^2 - b^2 = (a+b)(a-b)$$

完全平方公式:

$$a^2 + 2ab + b^2 = (a+b)^2$$

$$a^2 - 2ab + b^2 = (a-b)^2$$

立方和公式:

$$a^3 + b^3 = (a+b)(a^2 - ab + b^2)$$

立方差公式:

$$a^3 - b^3 = (a-b)(a^2 + ab + b^2)$$

例题 1 ▸

If $x + y = 2$ and $x^2 + y^2 = 2$, what is the value of xy?

(A) -2 (B) -1 (C) 0 (D) 1 (E) 2

解:

根据完全平方公式: $(x+y)^2 = x^2 + y^2 + 2xy$ 得到如下等式:

$$2^2 = 2 + 2xy,$$

则 $xy = 1$,答案为 D。

例题 2

Which of the following equations is NOT equivalent to $4x^2 = y^2 - 9$?

(A) $4x^2 + 9 = y^2$ (B) $4x^2 - y^2 = -9$ (C) $4x^2 = (y+3)(y-3)$

(D) $2x = y - 3$ (E) $x^2 = \dfrac{y^2 - 9}{4}$

解:

选项 D 等式两边平方后为 $4x^2 = (y-3)^2 = y^2 + 9 - 6y$。和题干中的算式不一致,故答案为 D。

例题 3

Is $x^2 - y^2$ a positive number?

(1) $x - y$ is a positive number.

(2) $x + y$ is a positive number.

(A) Statement (1) ALONE is sufficient, but statement (2) alone is not sufficient.

(B) Statement (2) ALONE is sufficient, but statement (1) alone is not sufficient.

(C) BOTH statements TOGETHER are sufficient, but NEITHER statement ALONE is sufficient.

(D) EACH statement ALONE is sufficient.

(E) Statements (1) and (2) TOGETHER are NOT sufficient.

解:

题目要我们判断 $(x+y)(x-y)$ 是否为正数。

显然单独使用两个条件是无法得出 $(x+y)(x-y)$ 的情况的。

同时使用两个条件可以知道 $(x+y)(x-y)$ 必然是正数,故条件 1 + 条件 2 充分。

综上,答案为 C。

例题 4

If $N = \dfrac{K}{T + \dfrac{x}{y}}$, where $T = \dfrac{K}{5}$ and $x = 5 - T$, which of the following expresses y in terms

of N and T?

(A) $\dfrac{N(5-T)}{T(5-N)}$ (B) $\dfrac{N(T-5)}{T(5-N)}$ (C) $\dfrac{5N(5-T)}{T(5-N)}$

(D) $\dfrac{5N(5-T)}{T(1-5N)}$ (E) $\dfrac{N(5-T)}{5}$

解:

先把 $T + \dfrac{x}{y}$ 和 N 对换位置，则有 $\dfrac{x}{y} = \dfrac{K}{N} - T$。

代入 $K = 5T$，则有：

$$\frac{x}{y} = \frac{5T}{N} - T,$$

$$y = \frac{5T - TN}{Nx}。$$

代入 $x = 5 - T$，则有：

$$y = \frac{N(5-T)}{T(5-N)}。$$

答案为 A。

例题 5

Which of the following equations is NOT equivalent to $4x^2 = y^2 - 9$?

(A) $4x^2 + 9 = y^2$ (B) $4x^2 - y^2 = -9$ (C) $4x^2 = (y+3)(y-3)$

(D) $2x = y - 3$ (E) $x^2 = \dfrac{y^2 - 9}{4}$

解:

选项 A，B 只需要移项即可得；选项 C 需要把 $y^2 - 9$ 利用平方差公式展开；选项 E 需要等式两边同时除以 4。只有选项 D 无法通过 $4x^2 = y^2 - 9$ 变化得出。故答案为 D。

例题 6

If $x \geq 0$ and $x = \sqrt{8xy - 16y^2}$, then, in terms of y, $x =$

（A）$-4y$ （B）$\dfrac{y}{4}$ （C）y （D）$4y$ （E）$4y^2$

解：

等式两边同时平方则有：

$$x^2 = 8xy - 16y^2,$$

即

$$x^2 - 8xy + 16y^2 = 0,$$
$$(x - 4y)^2 = 0,$$

则有：

$$x = 4y。$$

综上，答案为 D。

3.2 ▶ 方程

3.2.1 ▶ 一元一次方程

一元一次方程指只含有一个未知数、未知数的最高次数为 1 且两边都为整式的等式。一元一次方程只有一个根。一元一次方程可以解决绝大多数的工程问题、行程问题、分配问题、盈亏问题、积分表问题、电话计费问题和数字问题。

一元一次方程是初中一年级的必修内容，相信大家对"设 x"和"找方程的姐姐（解）"都不陌生。只要能设出 x，直接运算即可。

147

例题 1

The maximum recommended pulse rate R, when exercising, for a person who is x years of age is given by the equation $R = 176 - 0.8x$. What is the age, in years, of a person whose maximum recommended pulse rate when exercising is 140?

(A) 40　　　　(B) 45　　　　(C) 50　　　　(D) 55　　　　(E) 60

解:

本题可谓是送分送到家了, 连方程式都列好了。

$$R = 140 = 176 - 0.8x,$$

解得 $x = 45$, 答案为 B。

例题 2

The toll T, in dollars, for a truck using a certain bridge is given by the formula

$T = 1.50 + 0.50\ (x - 2)$, where x is the number of axles on the truck. What is the toll for an 18-wheel truck that has 2 wheels on its front axle and 4 wheels on each of its other axles?

(A) $2.50　　(B) $3.00　　(C) $3.50　　(D) $4.00　　(E) $5.00

解:

题目中给出了 cost 和 axles 的一元一次方程。想要计算费用, 就要知道卡车的 axles 数量。题目又给出了轮子的数量, 根据轮子的数量可以计算出 axles 的数量, 假设卡车有 x 个 axles, 则有:

$$2 \times 1 + 4 \times (x - 1) = 18,$$

解出 $x = 5$。此处要注意题目中说前轴有两个轮子, 其他每个轴有 4 个车轮, 所以其他轴的车轮总数为 $(x - 1) \times 4$。

将 $x = 5$ 代入 $T = 1.5 + 0.5(x - 2) = 1.5 + 0.5 \times 3 = 3$。答案为 B。

例题 3

If $k \neq 0$ and $k - \dfrac{3 - 2k^2}{k} = \dfrac{x}{k}$, then $x =$

(A) $-3 - k^2$ (B) $k^2 - 3$ (C) $3k^2 - 3$

(D) $k - 3 - 2k^2$ (E) $k - 3 + 2k^2$

解:

这道题我们可以把 k 看作已知数，等式两边同时乘 k，则有：

$$x = k^2 - (3 - 2k^2) = 3k^2 - 3。$$

答案为 C。

3.2.2 ▶ 二元一次方程

含有两个未知数，并且含有未知数的项的次数都是 1 的整式方程叫作二元一次方程，可简化为 $ax + by + c = 0$（a、$b \neq 0$）的一般式与 $ax + by = c$（a、$b \neq 0$）的标准式。

每个二元一次方程都有无数解，只有二元一次方程组才只有一组解。

例题 4

$$\begin{cases} 2x + 2y = -4 \\ 4x + y = 1 \end{cases}$$

In the system of equations above, what is the value of x?

(A) -3 (B) -1 (C) $\dfrac{2}{5}$ (D) 1 (E) $\dfrac{7}{4}$

解:

第二个方程可以转化成 $y = 1 - 4x$，代入第一个方程：$2x + 2(1 - 4x) = 2 - 6x = -4$。

解得 $x = 1$；$y = -3$。

答案为 D。

例题 5

The number of rooms at Hotel G is 10 less than twice the number of rooms at Hotel H. If the total number of rooms at Hotel G and Hotel H is 425, what is the number of rooms at Hotel G?

(A) 140　　　　(B) 180　　　　(C) 200　　　　(D) 240　　　　(E) 280

解:

设 Hotel H 的房间数量是 x, Hotel G 的房间数量是 y。根据题目中的两个条件可列二元一次方程组:

$$\begin{cases} y = 2x - 10; \\ x + y = 425。 \end{cases}$$

把 $y = 2x - 10$ 代入 $x + y = 425$, 可得 $x + 2x - 10 = 425$。

解出 $x = 145$; $y = 280$。

则 Hotel H 有 145 个房间, Hotel G 有 280 个房间。答案是 E。

例题 6

If $2x + 5y = 8$ and $3x = 2y$, what is the value of $2x + y$?

(A) 4　　　(B) $\dfrac{70}{19}$　　　(C) $\dfrac{64}{19}$　　　(D) $\dfrac{56}{19}$　　　(E) $\dfrac{40}{19}$

解:

依题意,

$$y = \frac{3}{2}x$$

$$2x + y = \frac{7}{2}x,$$

$$2x + 5y = \frac{19}{2}x = 8,$$

$$x = \frac{16}{19},$$

$$\frac{7}{2}x = \frac{56}{19},$$

答案是 D。

例题 7

There were 36,000 hardback copies of a certain novel sold before the paperback version was issued. From the time the first paperback copy was sold until the last copy of the novel was sold, 9 times as many paperback copies as hardback copies were sold. If a total of 441,000 copies of the novel were sold in all, how many paperback copies were sold?

(A) 45,000　　　　(B) 360,000　　　　(C) 364,500

(D) 392,000　　　　(E) 396,900

解：

依题意，设之后卖出的精装书册数为 X，卖出的平装书册数为 Y，

$$X + Y = 441000 - 36000 = 405000,$$

$$9X = Y,$$

解得 $Y = 364500$。

答案为 C。

3.2.3 ▶ 一元二次方程

只含有一个未知数（一元），并且未知数项的最高次数是 2（二次）的整式方程叫作一元二次方程。一元二次方程经过整理都可简化成一般形式 $ax^2 + bx + c = 0$（$a \neq 0$）。其中 ax^2 叫作二次项，a 是二次项系数；bx 叫作一次项，b 是一次项系数；c 叫作常数项。

一元二次方程主要考查求根公式、十字相乘法和韦达定理。

1 求根公式

$$x = \frac{-b \pm \sqrt{b^2 - 4ac}}{2a},$$

其中 $b^2 - 4ac$ 叫一元二次方程的判别式。

当 $b^2 - 4ac > 0$，方程有两个不相等的实数根。

当 $b^2 - 4ac = 0$，方程有两个相等的实数根。

当 $b^2 - 4ac < 0$，方程没有实数根。

❷ 十字相乘法

要想求得一个一元二次方程式的解，用求根公式自然是万能的，但往往计算十分复杂。要想又快又好地解题，十字相乘法就必须要学会。

十字相乘的方法简单来讲就是：十字左边相乘等于二次项，右边相乘等于常数项，交叉相乘再相加等于一次项。其实就是运用乘法公式运算来进行因式分解。例如用十字相乘的方式解出方程：$x^2 + 3x - 4 = 0$。

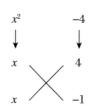

因为 $-x + 4x = 3x$，刚好等于一次项，十字成立。方程式可变为：

$$(x+4)(x-1) = 0,$$

可以解得 $x = -4$ 或 $x = 1$。

❸ 韦达定理

一元二次方程的两根之和为：

$$x_1 + x_2 = -\frac{b}{a}。$$

一元二次方程的两根之积为：

$$x_1 x_2 = \frac{c}{a}。$$

例题 8

What is the larger of the 2 solutions of the equation $x^2 - 4x = 96$?

(A) 8　　　　(B) 12　　　　(C) 16　　　　(D) 32　　　　(E) 100

解:

整理可得:

$$x^2 - 4x - 96 = 0,$$

十字相乘可得:

$$
\begin{array}{ccc}
x^2 & & -96 \\
\downarrow & & \downarrow \\
x & & -12 \\
& \times & \\
x & & 8
\end{array}
$$

原式可变为:

$$(x-12)(x+8) = 0,$$

解得: $x = 12$ 或 -8, 其中 $x = 12$ 为更大的解。

故答案为 B。

例题 9

In the equation $x^2 + bx + 12 = 0$, x is a variable and b is a constant. What is the value of b?

(1) $x - 3$ is a factor of $x^2 + bx + 12$.

(2) 4 is a root of the equation $x^2 + bx + 12 = 0$.

(A) Statement (1) ALONE is sufficient, but statement (2) alone is not sufficient.

(B) Statement (2) ALONE is sufficient, but statement (1) alone is not sufficient.

(C) BOTH statements TOGETHER are sufficient, but NEITHER statement ALONE is sufficient.

（D）EACH statement ALONE is sufficient.

（E）Statements（1）and（2）TOGETHER are NOT sufficient.

解：

条件 1 说，$x-3$ 是 $x^2+bx+12$ 的一个因子。因为 x^2 的系数为 1，所以设另一个因子为 $x+k$，则有：

$$(x-3)(x+k)=0,$$

展开后为：

$$x^2+(k-3)x-3k=0,$$

和原式 $x^2+bx+12$ 对应可得：$k=-4$；$b=-7$。故条件 1 充分。

条件 2 说，4 是 $x^2+bx+12$ 的一个根。将 $x=4$ 代入 $x^2+bx+12=0$，解得 $b=-7$，故条件 2 充分。

综上，答案为 D。

例题 10

If b, c and d are constants and $x^2+bx+c=(x+d)^2$ for all values of x, what is the value of c?

（1）$d=3$

（2）$b=6$

（A）Statement（1）ALONE is sufficient, but statement（2）alone is not sufficient.

（B）Statement（2）ALONE is sufficient, but statement（1）alone is not sufficient.

（C）BOTH statements TOGETHER are sufficient, but NEITHER statement ALONE is sufficient.

（D）EACH statement ALONE is sufficient.

（E）Statements（1）and（2）TOGETHER are NOT sufficient.

解：

首先可以将 $(x+d)^2$ 展开，则可得：

$$x^2 + bx + c = x^2 + 2dx + d^2,$$

由此可知，$c = d^2$；只需知道 d^2 即可。又由于 $b = 2d$，所以知道 b 的值就可以求 d 的值，自然也可以得知 d^2 的值。综上，答案为 D。

3.3 ▸ 不等式

3.3.1 ▸ 基础不等式

进行不等式的计算，最重要的是要掌握运算符号的规则和变化。

①不等式两边相加或相减同一个数或式子，不等号的方向不变。(移项要变号)

②不等式两边乘或除以同一个正数，不等号的方向不变。

③不等式两边乘或除以同一个负数，不等号的方向改变。

例题 1 ▸

If $-4 < x < 7$ and $-6 < y < 3$, which of the following specifies all the possible values of $x + y$?

(A) $-10 < x + y < 10$ (B) $-1 < x + y < 7$ (C) $-10 < x + y < 7$

(D) $-4 < x + y < 3$ (E) $-6 < x + y < 10$

解：

本题非常简单，不等式的左边和右边分别相加即可，答案为 A。

例题 2

If $x+y+z>0$, is $z>1$?

(1) $z>x+y+1$

(2) $x+y+1<0$

(A) Statement (1) ALONE is sufficient, but statement (2) alone is not sufficient.

(B) Statement (2) ALONE is sufficient, but statement (1) alone is not sufficient.

(C) BOTH statements TOGETHER are sufficient, but NEITHER statement ALONE is sufficient.

(D) EACH statement ALONE is sufficient.

(E) Statements (1) and (2) TOGETHER are NOT sufficient.

解:

条件 1,我们无法得知 $x+y$ 是否大于 0,所以无法得知 z 是否大于 1,故条件 1 不充分。

条件 2,如果 $x+y+1<0$,则 $x+y<-1$。又因为 $x+y+z>0$,所以此时 z 必然大于 -1。试想,z 加上比 -1 还小的数都要大于 0,z 肯定要大于 1,故条件 2 充分。

综上,答案为 B。

例题 3

If there is a least integer that satisfies the inequality $\frac{9}{x} \geqslant 2$, what is that least integer?

(A) 0　　　　(B) 1　　　　(C) 4　　　　(D) 5

(E) There is not a least integer that satisfies the inequality.

解:

若 $\frac{9}{x} \geqslant 2$,则此时 x 必然是大于 0 的,所以不等式两边同时乘 x,不等号不变向,即

$2x-9 \leqslant 0$ 且 $x \neq 0$,

解出：

$$0 < x \leqslant \frac{9}{2},$$

最小的整数必然为1。

综上，答案为B。

例题4

If S is the sum of the first n positive integers, what is the value of n?

（1）$S < 20$

（2）$S^2 > 220$

（A）Statement（1）ALONE is sufficient, but statement（2）alone is not sufficient.

（B）Statement（2）ALONE is sufficient, but statement（1）alone is not sufficient.

（C）BOTH statements TOGETHER are sufficient, but NEITHER statement ALONE is sufficient.

（D）EACH statement ALONE is sufficient.

（E）Statements（1）and（2）TOGETHER are NOT sufficient.

解：

条件1说，S小于20，显然有前2位、前3位等好几种情况满足这个条件，故条件1不充分。

条件2说，S的平方大于220。因为自然数没有上限，所以满足这个条件的n有无数个，故条件2不充分。

两个条件同时成立时，只有当$n = 5$时，$S = 15$且$S^2 = 225$满足条件，故条件1+条件2充分。

综上，答案为C。

Is $xy < 6$?

(1) $x < 3$ and $y < 2$

(2) $\dfrac{1}{2} < x < \dfrac{2}{3}$ and $y^2 < 64$

(A) Statement (1) ALONE is sufficient, but statement (2) alone is not sufficient.

(B) Statement (2) ALONE is sufficient, but statement (1) alone is not sufficient.

(C) BOTH statements TOGETHER are sufficient, but NEITHER statement ALONE is sufficient.

(D) EACH statement ALONE is sufficient.

(E) Statements (1) and (2) TOGETHER are NOT sufficient.

解:

条件 1 给出了 x 和 y 的范围。但是,如果 x 和 y 均为负数,例如,$x = -4$,$y = -7$,此时两者相乘的积必然大于 6,故条件 1 不充分。

由条件 2 可以解出 $-8 < y < 8$。因为 x 为正数,所以当 $y < 0$ 时,xy 必然小于 6;当 $y > 0$,x 取极大值 $\dfrac{2}{3}$ 且 y 取极大值 8 时,xy 为极大值,即 $xy < \dfrac{16}{3} < 6$,故条件 2 充分。

综上,答案为 B。

Is $-3 \leqslant x \leqslant 3$?

(1) $x^2 + y^2 = 9$

(2) $x^2 + y \leqslant 9$

(A) Statement (1) ALONE is sufficient, but statement (2) alone is not sufficient.

(B) Statement (2) ALONE is sufficient, but statement (1) alone is not sufficient.

(C) BOTH statements TOGETHER are sufficient, but NEITHER statement ALONE is sufficient.

(D) EACH statement ALONE is sufficient.

(E) Statements (1) and (2) TOGETHER are NOT sufficient.

解：

条件 1，y^2 的极小值为 0，所以 x^2 的极大值为 9，即 $x^2 \leq 9$。解出 $-3 \leq x \leq 3$，故条件 1 充分。

条件 2，$x^2 + y \leq 9$，此时 y 可以为负数，所以 x^2 的极值理论上可以是无限大，无法推理出 x 是否在 3 和 -3 之间。故条件 2 不充分。

综上，答案为 A。

例题 7

Is x less than y?

（1）$x - y + 1 < 0$

（2）$x - y - 1 < 0$

（A）Statement（1）ALONE is sufficient，but statement（2）alone is not sufficient.

（B）Statement（2）ALONE is sufficient，but statement（1）alone is not sufficient.

（C）BOTH statements TOGETHER are sufficient，but NEITHER statement ALONE is sufficient.

（D）EACH statement ALONE is sufficient.

（E）Statements（1）and（2）TOGETHER are NOT sufficient.

解：

由条件 1 得出 $x - y < -1$，这表明 x 减去 y 比 0 还小，证明 x 必然比 y 小，故条件 1 充分。

由条件 2 得出 $x - y < 1$，x 减去 y 比 1 小，无法证明 x 比 y 小，题目中并没有说 x 和 y 都是整数。故条件 2 不充分。

综上，答案为 A。

例题 8

If $x^2 - 2 < 0$，which of the following specifies all the possible values of x?

（A）$0 < x < 2$　　　　（B）$0 < x < \sqrt{2}$　　　　（C）$-\sqrt{2} < x < \sqrt{2}$

（D）　$-2<x<0$　　　　（E）　$-2<x<2$

解：

移项可得：

$$x^2<2。$$

因为无论正负数，平方后都为正数，所以 x^2 开方后的范围是：

$$-\sqrt{2}<x<\sqrt{2}。$$

答案为 C。

3.3.2 ▶ 极值问题

问"可能值"的考题绝大部分实质上都是极值问题。例如，题目问：

如果 $x<2$，以下哪项是 $x+2$ 的可能值？

类似这样的问题，我们应该先求出 $x+2$ 的极限为 4。由此可知，在极限范围内的数字都是 $x+2$ 的可能值。

因此，考试中但凡看到可能值问题，均可以先求出极值，然后再考虑可能值。

例题 9

If the average（arithmetic mean）of positive integers x，y，and z is 10，what is the greatest possible value of z？

（A）8　　　　（B）10　　　　（C）20　　　　（D）28　　　　（E）30

解：

依题意，$x+y+z=30$。要 z 取最大值，必然是 x 和 y 取最小值。由于 x 和 y 都是正整数，所以它们的最小值为 1。因此，z 的极大值为 28。答案为 D。

例题 10

Five pieces of wood have an average (arithmetic mean) length of 124 centimeters and a median length of 140 centimeters. What is the maximum possible length, in centimeters, of the shortest piece of wood?

(A) 90   (B) 100   (C) 110   (D) 130   (E) 140

解:

依题意，因为五块木头的长度中位数为 140，所以中间长度的木头长度必为 140。由于平均数固定，所以要想让最短的木头最长，一定是让比较长的木头尽量短。

因此，假设长度大于或等于中位数的两块木头的长度均为 140，则 $124 \times 5 - 140 \times 3 = 200$。

由此可知，两块长度短于或等于中位数的木头长度之和为 200。因此，最短的木头的最大长度为 100。答案为 B。

例题 11

A certain city with a population of 132,000 is to be divided into 11 voting districts, and no district is to have a population that is more than 10 percent greater than the population of any other district. What is the minimum possible population that the least populated district could have?

(A) 10,700   (B) 10,800   (C) 10,900   (D) 11,000   (E) 11,100

解:

题目的意思是，将 132000 分成 11 份，且最小的和最大的相差不能超过 10%。

试想一个情景，10 个区将人口少的机会都给其中一个，10 个超过 12000，只有 1 个低于 12000（$12000 = \dfrac{132000}{11}$，是把所有人口纯粹平均地分给 11 个区后的数值）。

那么，人口最少的区应该满足：其他区人数相等，并且是它的 1.1 倍。

$$X + 10Y = 132000,$$

$$1.1X = Y。$$

解得 $X = 11000$。答案为 D。

例题 12

List T consist of 30 positive decimals, none of which is an integer, and the sum of the 30 decimals is S. The estimated sum of the 30 decimals, E, is defined as follows. Each decimal in T whose tenths digit is even is rounded up to the nearest integer, and each decimal in T whose tenths digits is odd is rounded down to the nearest integer. If $\frac{1}{3}$ of the decimals in T have a tenths digit that is even, which of the following is a possible value of $E - S$?

Ⅰ. -16 Ⅱ. 6 Ⅲ. 10

（A）Ⅰ only   （B）Ⅰ and Ⅱ only   （C）Ⅰ and Ⅲ only

（D）Ⅱ and Ⅲ only   （E）Ⅰ，Ⅱ and Ⅲ

解：

本题先要理解题目的意思。T 中有 30 个数（都不是整数），它们的和是 S；E 的定义是，但凡小数点后 1 位是偶数，则个位数 $+1$；但凡小数点后 1 位是奇数，则个位数不变。例如，如果是 11.2，则在 E 中它等于 12；如果是 11.9，则在 E 中它等于 11。题目问的是哪个值是 $E - S$ 的可能值。

我们依然要先找到极限情况。

E 中十分位是偶数的数字最多和 S 中十分位是偶数的数字相差 0.8。一共有 10 个十分位是偶数的数字，所以总体最多相差 8；反之，E 中十分位是偶数的数字最少和 S 中十分位是偶数的数字相差 0.2，总体最多相差 2。

E 中十分位是奇数的数字最多和 S 中十分位是奇数的数字相差 -0.9。一共有 20 个十分位是奇数的数字，所以总体最多差 -18；反之，E 中十分位是奇数的数字最少和 S 中十分位是奇数的数字相差 -0.1，总体最多相差 -2。

由此可知，$E - S$ 的取值范围应为 -16 到 6（$-16 < E - S < 6$）。因此，答案为 B。

3.3.3 ▶ 一元高次不等式

解一元高次不等式是我们初高中学习过的概念，但应该有不少考生都忘光了。下面就让我们一起来复习一下解题要求。

一元高次不等式的解法是数轴穿根法。方法如下：

①将不等式的首项系数化为"正"。

②移项通分，不等号右侧化为"0"。

③因式分解，化为几个一次因式积的形式，假设不等式为等式后求根。

④在数轴上把几个根都标出来。

⑤用线穿根。从最大值的右上方开始向左画线，每经过一个根，就穿过一次数轴，依次穿过。若不等号为 >，则取数轴上方部分；若不等号为 <，则取数轴下方部分。

例题 13

If $(x+2)(x-1)(x-4)>0$, $x=$

（A）$x<-2$ or $1<x<4$ （B）$-2<x<1$ or $x>4$ （C）$-2<x<4$

（D）$-2<x<1$ or $1<x<4$ （E）$x<-2$ or $1<x<4$

解：

利用穿根法画图，可得：

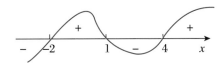

因为不等式是大于 0 的，因此取数轴上方区域，答案为 B。

3.3.4 ▸ 绝对值不等式

顾名思义，绝对值不等式就是不等式中会出现绝对值的情况。求解这种不等式需分类讨论去掉绝对值，然后根据不同的情况分别求解，最后取并集得出最终答案。

例如 $|3x-1|<2$，可以变为：$-2<3x-1<2$；两端分别求解可得 $-\dfrac{1}{3}<x<1$。

$3r \le 4s + 5$

$|s| \le 5$

Given the inequalities above, which of the following CANNOT be the value of r?

(A) -20 (B) -5 (C) 0 (D) 5 (E) 20

解：

$|s| \le 5$ 表明 $-5 \le s \le 5$，即 s 的最小值是 -5，最大值是 5。将最大值代入 $3r \le 4s + 5$，则有：$3r \le 4 \times 5 + 5$；$r \le \dfrac{25}{3}$。显然，只有 20 不满足条件，故答案为 E。

$2x + y = 12$

$|y| \le 12$

For how many ordered pairs (x, y) that are solutions of the system above are x and y both integers?

(A) 7 (B) 10 (C) 12 (D) 13 (E) 14

解：

依题意可知：

$$-12 \le y \le 12。$$

题目要求 x 和 y 必须都是整数，因此，我们需要把 y 的所有整数值都代入 $2x + y = 12$ 中验证。

请注意，当 y 是奇数时，$12 - y$ 必然是奇数。奇数除以 2 后必然不是整数，所以 y 是奇数的时候 x 必然不为整数。当 y 是偶数时，因为偶数必然至少含有一个 2，所以 $12 - y$ 除以 2 后必为整数。因此，y 中所有的偶数都是满足题设的。从 -12 到 12 中，一共有 12 个偶数。此时不要忘了，y 还可以等于 0，0 也是整数。因此，一共有 13 组解。答案为 D。

3.4 ▶ 数列

3.4.1 ▶ 等差和等比数列

原则上来说，数列就是一列数。有两个常考的特殊数列——等差数列和等比数列。

等差数列是指从第二项起，每一项与它的前一项的差等于同一个常数的一种数列。

通项公式为：

$$a_n = a_1 + (n-1) \times d。$$

前 n 项的求和公式为：

$$S_n = \frac{n(a_1 + a_n)}{2}。$$

或者可以写为：

$$S_n = na_1 + \frac{n(n-1)}{2}d。$$

其中 a_1 为首项，d 是公差，a_n 是末项。

等比数列是指从第二项起，每一项与它的前一项的比值等于同一个常数的一种数列。

通项公式为：

$$a_n = a_1 \cdot q^{n-1}。$$

前 n 项的求和公式为：

$$S_n = na_1 \quad (q=1)，$$

$$S_n = \frac{a_1(1-q^n)}{1-q} \quad (q \neq 1)。$$

有关数列的考题绝大部分都是直接计算题，相对来说十分简单。也有许多考题虽然涉及数列，但完全不需要用到等差或等比数列的公式。

例题 1

If the sum of 7 consecutive integers is 434, what is the greatest integer of these integers?

(A) 59   (B) 61   (C) 65   (D) 67   (E) 69

解:

连续正整数是公差为 1 的等差数列,代入求和公式则有:

$$434 = 7a_1 + \frac{7 \times (7-1)}{2},$$

解得 $a_1 = 59$,最大值为 $59 + 6 = 65$。

例题 2

$2 + 2 + 2^2 + 2^3 + 2^4 + 2^5 + 2^6 + 2^7 + 2^8 =$

(A) 2^9 (B) 2^{10} (C) 2^{16} (D) 2^{35} (E) 2^{37}

解:

显然,数列从第二项开始为首项是 2、公比是 2 的等比数列。其和为:

$$\frac{2 \times (1 - 2^8)}{1 - 2} = -2 + 2^9,$$

计算总和后别忘了第一项,整体的和为 2^9,答案为 A。

例题 3

The sequence a_1, a_2, $\cdots$, a_n, $\cdots$ is such that $a_n = 2a_{n-1} - x$ for all positive integers $n > 2$ and for a certain number x. If $a_5 = 99$ and $a_3 = 27$, what is the value of x?

(A) 3   (B) 9   (C) 18   (D) 36   (E) 45

解:

$$a_5 = 2a_4 - x = 99,$$

$$a_4 = 2a_3 - x = 2 \times 27 - x = 54 - x,$$

$$a_5 = 2 \times (54 - x) - x = 99,$$

$$108 - 2x - x = 99,$$
$$-3x = -9, \quad x = 3_{\circ}$$

答案为 A。

例题 4

The infinite sequence a_1, a_2, $\cdots$, a_n, $\cdots$ is such that $a_1 = 2$, $a_2 = -3$, $a_3 = 5$, $a_4 = -1$, and $a_n = a_{n-4}$ for $n > 4$. What is the sum of the first 97 terms of the sequence?

(A) 72   (B) 74   (C) 75   (D) 78   (E) 80

解：

$a_5 = a_{5-4} = a_1 = 2$；$a_6 = a_{6-4} = a_2 = -3$；$a_7 = a_{7-4} = a_3 = 5$；$a_8 = a_{8-4} = a_4 = -1$，

以此类推，每四个为一组，和是 3。从 a_1 到 a_{97}，共有 24 组完整的，另余下 1 个。因此前 97 项之和为 $24 \times 3 + a_{97}$，即 $24 \times 3 + 2 = 74$。

答案为 B。

例题 5

In the sequence 1，2，4，8，16，32，$\cdots$，each term after the first is twice the previous term. What is the sum of the 16th，17th，and 18th terms in the sequence?

(A) 2^{18} (B) $3 \ (2^{17})$ (C) $7 \ (2^{16})$

(D) $3 \ (2^{16})$ (E) $7 \ (2^{15})$

解：

题干中给定的数列很明显是一个等比数列。通项公式为 $a_n = 1 \times 2^{n-1} = 2^{n-1}$。

第 16 项为 2^{15}；第 17 项为 2^{16}；第 18 项为 2^{17}。

三者相加为 7×2^{15}，答案为 E。

例题 6

The sequence a_1, a_2, a_3, $\cdots$, a_n of n integers is such that $a_k = k$ if k is odd and $a_k = -a_{k-1}$ if k is even. Is the sum of the terms in the sequence positive?

(1) n is odd.

(2) a_n is positive.

(A) Statement (1) ALONE is sufficient, but statement (2) alone is not sufficient.

(B) Statement (2) ALONE is sufficient, but statement (1) alone is not sufficient.

(C) BOTH statements TOGETHER are sufficient, but NEITHER statement ALONE is sufficient.

(D) EACH statement ALONE is sufficient.

(E) Statements (1) and (2) TOGETHER are NOT sufficient.

解：

题干的意思是：当 k 是奇数时，$a_k = k$，例如 $a_1 = 3$，$a_3 = 3$；当 k 是偶数时，$a_k = -a_{k-1}$，例如 $a_2 = -1$，$a_4 = -3$。问所有项的和是否是正数。

条件 1 说，n 是奇数。如果 n 是奇数，则说明数列最后一个数一定是正数，且前面所有项之和为 0。因此，条件 1 充分。

条件 2 说，最后一项是正数。这一点说明最后一项 n 一定是奇数，情况和条件 1 相同，故条件 2 充分。

综上，答案为 D。

例题 7

In the finite sequence of positive integers K_1, K_2, K_3, $\cdots$, K_9, each term after the second is the sum of the two terms immediately preceding it. If $K_5 = 18$, what is the value of K_9?

(1) $K_4 = 11$

(2) $K_6 = 29$

(A) Statement (1) ALONE is sufficient, but statement (2) alone is not sufficient.

(B) Statement (2) ALONE is sufficient, but statement (1) alone is not sufficient.

（C）BOTH statements TOGETHER are sufficient, but NEITHER statement ALONE is sufficient.

（D）EACH statement ALONE is sufficient.

（E）Statements（1）and（2）TOGETHER are NOT sufficient.

解：

题干问：K_9 等于几？

依题意，$K_9 = K_7 + K_8 = K_5 + K_6 + K_6 + K_7 = K_5 + K_6 + K_6 + K_5 + K_6 = 2K_5 + 3K_6$。

条件 1 说，$K_4 = 11$，只要知道 K_4，就可以知道 K_6，故条件 1 充分。

条件 2 说，$K_6 = 29$，显然条件 2 也是充分的。

综上，答案为 D。

3.4.2 ▶ 递推数列

除了等差和等比数列外，GMAT 数学还经常考查考生对于递推数列的理解。同时，递推数列的考题一定会给出一个递推公式。

如果数列 a_n 的第 n 项与它前一项或几项的关系可以用一个式子来表示，那么这个公式叫作这个数列的递推公式。

我们直接来看一道例题：

有一个数列 a_n，它满足 $a_1 = 3$，$a_{n+1} = 2a_n + 1$，问 a_n 的通项公式。

考题中，$a_{n+1} = 2a_n + 1$ 就是一个递推公式。通过这个递推公式可知，数列 a_n 既不是等差数列，也不是等比数列。而我们只能写出等差或等比的通项公式。所以可以确定以下的解题思路和方向，即要想办法基于 a_n 构造出一个等差或等比数列，将其转换成我们掌握的形式，进而写出数列通项。

由于第 $n+1$ 项是第 n 项的 2 倍，所以可以尝试让 a_n 中的每一项都加上或者减去一个常数以构造一个新的等比数列。设这个数为 x，则递推公式为：

$$a_{n+1} + x = 2(a_n + x)。$$

让这个递推公式与题干中的递推公式相等，确认 x 的值。

$$x = 1,$$

递推公式可以改写为：

$$a_{n+1} + 1 = 2\left(a_n + 1\right),$$

设 $a_n + 1 = b_n$，则 $b_{n+1} = a_{n+1} + 1$；$b_1 = 4$，

将 b_n 代入原递推公式，则有

$$b_{n+1} = 2b_n,$$

显然，b_n 是首项为 4，公比为 2 的等比数列。

直接写出 b_n 的通项公式：

$$b_n = b_1 \times q^{n-1} = 4 \times 2^{n-1},$$
$$a_n = b_n - 1 = 2^{n+1} - 1。$$

于是 a_n 的通项公式为：

GMAT 数学基本上只会考查这种可以转化为等比数列的情况，所以大家只需牢记转化过程即可成功完成考题。

例题 8

If the sequence x_1, x_2, x_3, $\cdots$, x_n is such that $x_1 = 3$ and $x_{n+1} = 2x_n - 1$ for $n \geq 1$, then $x_{20} - x_{19} =$

（A）2^{19}　　　（B）2^{20}　　　（C）2^{21}　　　（D）$2^{20} - 1$　　　（E）$2^{21} - 1$

解：

设 $x_{n+1} + k = 2\left(x_n + k\right)$，

根据该式与题干中的递推公式相等，则有：

$$k = -1,$$

即递推公式可改为：

$$x_{n+1} - 1 = 2\left(x_n - 1\right),$$

设 $x_n - 1$ 为 b_n，则 $x_{n+1} - 1$ 为 b_{n+1}；$b_1 = 2$，

新数列 b_n 的通项公式为:

$$b_n = b_1 \times q^{n-1} = 2 \times 2^{n-1},$$

$$x_n = b_n + 1 = 2^n + 1$$

$$x_{20} - x_{19} = 2^{20} + 1 - 2^{19} - 1 = 2^{19},$$

答案为 A。

例题 9

$a_{n+1} = \sqrt{16a_n^2 + 9}$, 求 $\dfrac{a_{20}}{a_{19}} \approx$

(A) 2　　　　(B) 4　　　　(C) 6　　　　(D) 8　　　　(E) 10

解:

等式两端同时平方,则有:

$$a_{n+1}^2 = 16a_n^2 + 9,$$

设 $a_{n+1}^2 + x = 16(a_n^2 + x)$,

解得:

$$x = \frac{3}{5},$$

$$a_{n+1}^2 + \frac{3}{5} = 16\left(a_n^2 + \frac{3}{5}\right),$$

设 $b_n = a_n^2 + \dfrac{3}{5}$; 则 $b_{n+1} = a_{n+1}^2 + \dfrac{3}{5}$,

此时 b_n 为公比为 16 的等比数列。

$$\frac{b_{n+1}}{b_n} = \frac{a_{n+1}^2 + \dfrac{3}{5}}{a_n^2 + \dfrac{3}{5}} = 16,$$

由此可知, $\dfrac{a_{n+1}}{a_n} \approx 4$ (因为分子和分母都加上小于 1 的数,所以不太影响比值)。

综上,答案为 B。

例题 10

In the sequence of nonzero numbers t_1, t_2, t_3, $\cdots$, t_n; $t_{n+1} = \dfrac{t_n}{2}$ for all positive integers n. What is the value of t_5?

(1) $t_3 = \dfrac{1}{4}$ (2) $t_1 - t_5 = \dfrac{15}{16}$

(A) Statement (1) ALONE is sufficient, but statement (2) alone is not sufficient.

(B) Statement (2) ALONE is sufficient, but statement (1) alone is not sufficient.

(C) BOTH statements TOGETHER are sufficient, but NEITHER statement ALONE is sufficient.

(D) EACH statement ALONE is sufficient.

(E) Statements (1) and (2) TOGETHER are NOT sufficient.

解:

从题干给出的递推公式可以看出，t_n 是一个公比为 $\dfrac{1}{2}$ 的等比数列。所以，只要能知道该数列中某一项的值，就可以求得第五项的值。

条件 1 给出了第三项的值，显然可以根据等比数列的通项公式求出第五项的值，故条件 1 充分。

条件 2 给出了第一项和第五项的差，通过等比数列的通项公式和公比 $\dfrac{1}{2}$ 可以求出 t_1 的值，进而一定可以求出 t_5 的值，故条件 2 充分。

综上，答案为 D。

3.5 ▶ 函数

函数在 GMAT 数学中非常简单，可分为两种：一种是函数自变量替换，一种是定义新函数。

3.5.1 ▶ 函数自变量替换

所谓自变量替换，指的是将函数中的自变量统一替换为指定字母或字母表达式，例如：

$$f(x) = 2x^2 + 2。$$

而 $f(x-3)$ 的意思是将上述表达式中所有的 x 均替换为 $x-3$，则有：

$$f(x-3) = 2(x-3)^2 + 2。$$

例题 1 ▷

For which of the following functions f is $f(x) = f(1-x)$ for all x?

(A) $f(x) = 1-x$　　　(B) $f(x) = 1-x^2$　　　(C) $f(x) = x^2 - (1-x)^2$

(D) $f(x) = x^2(1-x)^2$　　(E) $f(x) = \dfrac{x}{1-x}$

解：

理解清楚题意后本题十分容易作答。只要把 $f(x)$ 中的自变量 x 替换为 $1-x$，整个函数依然和以前相同即可。显然，只有选项 D，将所有的 x 换为 $1-x$ 后，值不变，即 $(1-x)^2 x^2$。答案为 D。

例题 2 ▷

The function f is defined by $f(x) = -\dfrac{1}{x}$ for all nonzero numbers x. If $f(a) = -\dfrac{1}{2}$ and $f(ab) = \dfrac{1}{6}$, then $b =$

(A) 3　　　(B) $\dfrac{1}{3}$　　　(C) $-\dfrac{1}{3}$　　　(D) -3　　　(E) -12

解：

$$f(a) = -\frac{1}{a} = -\frac{1}{2},$$

因此，

$$a = 2,$$

$$f(2 \times b) = -\frac{1}{2 \times b} = \frac{1}{6}。$$

即

$$-2b = 6；\quad b = -3。$$

答案为 D。

例题 3

The function f is defined by $f(x) = \sqrt{x} - 10$ for all positive numbers x. If $u = f(t)$ for some positive numbers t and u. What is t in terms of u?

(A) $\sqrt{u+10}$　　　(B) $\left(\sqrt{u}+10\right)^2$　　　(C) $\sqrt{u^2+10}$

(D) $\left(u+10\right)^2$　　　(E) $\left(u^2+10\right)^2$

解:

将自变量 t 代入 $f(x) = \sqrt{x} - 10$, 则有:

$$f(t) = \sqrt{t} - 10$$

因为 $f(t) = u$, 所以 $\sqrt{t} - 10 = u$。

$$t = \left(u+10\right)^2。$$

答案为 D。

3.5.2 ▶ 定义新函数

所谓定义新函数, 指的是题干中给出一个特殊符号或者函数, 要求我们利用该函数的信息解题。这类考题虽然看起来吓人, 但仔细阅读后会发现都很简单, 属于基本应用。

例题 4

The function f is defined for each positive three-digit integer n by $f(n) = 2^x 3^y 5^z$, where x, y, and z are the hundreds, tens, and units digits of n, respectively. If m and v are three-digit positive integers such that $f(m) = 9f(v)$, then $m - v =$

(A) 8　　　(B) 9　　　(C) 18　　　(D) 20　　　(E) 80

解:

本题的题意较难理解, 我们先翻译一下:

定义一个函数, 函数的变量 n 是一个三位数正整数, 函数的表达式为:

$f(n) = 2^x \times 3^y \times 5^z$，$x$，$y$，$z$ 分别是 n 的百位、十位和个位数。问如果 m 和 v 分别是两个三位数正整数，还有 $f(m) = 9f(v)$，那么 $m - v$ 等于多少？

翻译清楚了考题，这个题就非常简单了。因为 $f(m) = 9f(v)$，而 $9 = 3^2$，这正好与 $f(n)$ 中的 3^y 对应，所以 m 一定是在十位比 v 多 2，因此，$m - v = 20$。答案为 D。

例题 5

For all numbers s and t, the operation $*$ is defined by $s * t = (s-1)(t+1)$. If $(-2) * x = -12$, then $x =$

(A) 2   (B) 3   (C) 5   (D) 6   (E) 11

解：

只要能看懂 $*$ 的算法即可。代入 $*$ 的算法，则有：

$$(-2) * x = (-2-1)(x+1) = -12,$$
$$-3x - 3 = -12,$$
$$-3x = -9,$$
$$x = 3。$$

答案为 B。

例题 6

If @ denotes a mathematical operation, does $x@y = y@x$ for all x and y?

(1) For all x and y, $x@y = 2(x^2 + y^2)$.

(2) For all y, $0@y = 2y^2$.

(A) Statement (1) ALONE is sufficient, but statement (2) alone is not sufficient.

(B) Statement (2) ALONE is sufficient, but statement (1) alone is not sufficient.

(C) BOTH statements TOGETHER are sufficient, but NEITHER statement ALONE is sufficient.

（D）EACH statement ALONE is sufficient.

（E）Statements（1）and（2）TOGETHER are NOT sufficient.

解：

题目问的是@函数是否有交换规律。

条件1，将其中的 x 和 y 交换，则有：

$$y@x = 2(y^2 + x^2),$$

其显然等同于 $2(x^2 + y^2)$。故条件1充分。

条件2，只知道0和 y 经过@函数后的结果，无法确定 $x@y$ 是否有交换规律（x 不一定等于0）。故条件2不充分。

综上，答案为A。

例题7

If the symbol ▼ represents either addition，subtraction，multiplication，or division，what is the value of 6 ▼ 2?

（1）10 ▼ 5 = 2

（2）4 ▼ 2 = 2

（A）Statement（1）ALONE is sufficient，but statement（2）alone is not sufficient.

（B）Statement（2）ALONE is sufficient，but statement（1）alone is not sufficient.

（C）BOTH statements TOGETHER are sufficient，but NEITHER statement ALONE is sufficient.

（D）EACH statement ALONE is sufficient.

（E）Statements（1）and（2）TOGETHER are NOT sufficient.

解：

题目问的是我们能否确认▼究竟是加减乘除的哪一种。

条件1，仅当▼表示除法时，10 ▼ 5 = 2 才成立，故条件1充分。

条件2，无论▼是减法还是除法，4 ▼ 2 = 2 均成立，因此无法确定▼，故条件2不充分。

综上，答案为A。

例题 8

For any integer k greater than 1, the symbol k^* denotes the product of all the fractions of the form $\frac{1}{t}$, where t is an integer between 1 and k, inclusive. What is the value of $\frac{5^*}{4^*}$?

(A) 5   (B) $\frac{5}{4}$ (C) $\frac{4}{5}$ (D) $\frac{1}{4}$ (E) $\frac{1}{5}$

解:

$$5^* = 1 \times \frac{1}{2} \times \frac{1}{3} \times \frac{1}{4} \times \frac{1}{5},$$

$$4^* = 1 \times \frac{1}{2} \times \frac{1}{3} \times \frac{1}{4},$$

$$\frac{5^*}{4^*} = \frac{1}{5} \text{。}$$

答案为 E。

例题 9

The operation Δ is defined for all nonzero x and y by $x \Delta y = x + \frac{x}{y}$. If $a > 0$, then $1 \Delta (1 \Delta a) =$

(A) a (B) $a + 1$ (C) $\frac{a}{a+1}$

(D) $\frac{a+2}{a+1}$ (E) $\frac{2a+1}{a+1}$

解:

$$1 \Delta a = 1 + \frac{1}{a}$$

$$1 \Delta \left(1 + \frac{1}{a}\right) = 1 + \frac{1}{1 + \frac{1}{a}} = 1 + \frac{a}{a+1} = \frac{2a+1}{a+1}$$

答案为 E。

1. The sequence a_1, a_2, a_3, a_4, a_5 is such that $a_n = a_{n-1} + 5$ for $2 \leqslant n \leqslant 5$. If $a_5 = 31$, what is the value of a_1?

(A) 1　　　(B) 6　　　(C) 11　　　(D) 16　　　(E) 21

2. After 4,000 gallons of water were added to a large water tank that was already filled to $\frac{3}{4}$ of its capacity, the tank was then at $\frac{4}{5}$ of its capacity. How many gallons of water does the tank hold when filled to capacity?

(A) 5,000　　(B) 6,200　　(C) 20,000　　(D) 40,000　　(E) 80,000

3. A certain drive-in movie theater has a total of 17 rows of parking spaces. There are 20 parking spaces in the first row and 21 parking spaces in the second row. In each subsequent row there are 2 more parking spaces than in the previous row. What is the total number of parking spaces in the movie theater?

(A) 412　　(B) 544　　(C) 596　　(D) 632　　(E) 692

4. In a certain sequence, each term after the first term is one-half the previous term. If the tenth term of the sequence is between 0.0001 and 0.001, then the twelfth term of the sequence is between

(A) 0.0025 and 0.025　　　　　　(B) 0.00025 and 0.0025

(C) 0.000025 and 0.00025　　　　(D) 0.0000025 and 0.000025

(E) 0.00000025 and 0.0000025

5. What is the sum of the odd integers from 35 to 85, inclusive?

(A) 1,560　　(B) 1,500　　(C) 1,240　　(D) 1,120　　(E) 1,100

6. For all positive integers m and v, the expression $m \ominus v$ represents the remainder when m is divided by v. What is the value of $[(98 \ominus 33) \ominus 17] - [98 \ominus (33 \ominus 17)]$?

(A) -10　　(B) -2　　(C) 8　　(D) 13　　(E) 17

7. The "prime sum" of an integer n greater than 1 is the sum of all the prime factors of n, including repetitions. For example, the prime sum of 12 is 7, since $12 = 2 \times 2 \times 3$ and $2 + 2 + 3 = 7$. For which of the following integers is the prime sum greater than 35?

(A) 440　　(B) 512　　(C) 620　　(D) 700　　(E) 750

8. If $x^2 - 2x - 15 = 0$ and $x > 0$, which of the following must be equal to 0 ?

I. $x^2 - 6x + 9$

II. $x^2 - 7x + 10$

III. $x^2 - 10x + 25$

(A) I only　　　　　　(B) II only　　　　　　(C) III only

(D) II and III only　　(E) I, II, and III

9. If $x^2 + y^2 = 29$, what is the value of $(x + y)^2$?

(1) $xy = 10$

(2) $x = 5$

(A) Statement (1) ALONE is sufficient, but statement (2) alone is not sufficient.

(B) Statement (2) ALONE is sufficient, but statement (1) alone is not sufficient.

(C) BOTH statements TOGETHER are sufficient, but NEITHER statement ALONE is sufficient.

(D) EACH statement ALONE is sufficient.

(E) Statements (1) and (2) TOGETHER are NOT sufficient.

10. In the equation $x^2 + bx + 12 = 0$, x is a variable and b is a constant. What is the value of b?

(1) $x - 3$ is a factor of $x^2 + bx + 12$.

(2) 4 is a root of the equation $x^2 + bx + 12 = 0$.

(A) Statement (1) ALONE is sufficient, but statement (2) alone is not sufficient.

(B) Statement (2) ALONE is sufficient, but statement (1) alone is not sufficient.

(C) BOTH statements TOGETHER are sufficient, but NEITHER statement ALONE is sufficient.

(D) EACH statement ALONE is sufficient.

(E) Statements (1) and (2) TOGETHER are NOT sufficient.

11. If n is positive, is $\sqrt{n} > 100$?

   (1) $\sqrt{n-1} > 99$

   (2) $\sqrt{n+1} > 101$

   (A) Statement (1) ALONE is sufficient, but statement (2) alone is not sufficient.

   (B) Statement (2) ALONE is sufficient, but statement (1) alone is not sufficient.

   (C) BOTH statements TOGETHER are sufficient, but NEITHER statement ALONE is sufficient.

   (D) EACH statement ALONE is sufficient.

   (E) Statements (1) and (2) TOGETHER are NOT sufficient.

12. What is the sum of the first four terms of sequence S?

   (1) After the first two terms of S, the value of each term of S is equal to the average (arithmetic mean) of the last two preceding terms.

   (2) The average (arithmetic mean) of the first three terms of S is 10.

   (A) Statement (1) ALONE is sufficient, but statement (2) alone is not sufficient.

   (B) Statement (2) ALONE is sufficient, but statement (1) alone is not sufficient.

   (C) BOTH statements TOGETHER are sufficient, but NEITHER statement ALONE is sufficient.

   (D) EACH statement ALONE is sufficient.

   (E) Statements (1) and (2) TOGETHER are NOT sufficient.

13. A candle company determines that, for a certain specialty candle, the supply function is $p = m_1 x + b_1$ and the demand function is $p = m_2 x + b_2$, where p is the price of each candle, x is the number of candles supplied or demanded, and m_1, m_2, b_1, and b_1 are constants. At what value of x do the graphs of the supply function and demand function intersect?

   (1) $m_1 = -m_2 = 0.005$

   (2) $b_2 - b_1 = 6$

(A) Statement (1) ALONE is sufficient, but statement (2) alone is not sufficient.

(B) Statement (2) ALONE is sufficient, but statement (1) alone is not sufficient.

(C) BOTH statements TOGETHER are sufficient, but NEITHER statement ALONE is sufficient.

(D) EACH statement ALONE is sufficient.

(E) Statements (1) and (2) TOGETHER are NOT sufficient.

$14.$ For all x, the expression x^* is defined to be $ax + a$, where a is a constant. What is the value of 2^*?

(1) $3^* = 2$

(2) $5^* = 3$

(A) Statement (1) ALONE is sufficient, but statement (2) alone is not sufficient.

(B) Statement (2) ALONE is sufficient, but statement (1) alone is not sufficient.

(C) BOTH statements TOGETHER are sufficient, but NEITHER statement ALONE is sufficient.

(D) EACH statement ALONE is sufficient.

(E) Statements (1) and (2) TOGETHER are NOT sufficient.

$15.$ $[y]$ denotes the greatest integer less than or equal to y. Is $d < 1$?

(1) $d = y - [y]$

(2) $[d] = 0$

(A) Statement (1) ALONE is sufficient, but statement (2) alone is not sufficient.

(B) Statement (2) ALONE is sufficient, but statement (1) alone is not sufficient.

(C) BOTH statements TOGETHER are sufficient, but NEITHER statement ALONE is sufficient.

(D) EACH statement ALONE is sufficient.

(E) Statements (1) and (2) TOGETHER are NOT sufficient.

$16.$ The sequence s_1, s_2, s_3, ..., s_n is such that $s_n = \dfrac{1}{n} - \dfrac{1}{n+1}$ for all integers $n \geqslant 1$. If k is a positive integer, is the sum of the first k terms of the sequence greater than $\dfrac{9}{10}$?

(1) $k > 10$

(2) $k < 19$

(A) Statement (1) ALONE is sufficient, but statement (2) alone is not sufficient.

(B) Statement (2) ALONE is sufficient, but statement (1) alone is not sufficient.

(C) BOTH statements TOGETHER are sufficient, but NEITHER statement ALONE is sufficient.

(D) EACH statement ALONE is sufficient.

(E) Statements (1) and (2) TOGETHER are NOT sufficient.

17. In the sequence S of numbers, each term after the first two terms is the sum of the two immediately preceding terms. What is the 5th term of S?

(1) The 6th term of S minus the 4th term equals 5.

(2) The 6th term of S plus the 7th term equals 21.

(A) Statement (1) ALONE is sufficient, but statement (2) alone is not sufficient.

(B) Statement (2) ALONE is sufficient, but statement (1) alone is not sufficient.

(C) BOTH statements TOGETHER are sufficient, but NEITHER statement ALONE is sufficient.

(D) EACH statement ALONE is sufficient.

(E) Statements (1) and (2) TOGETHER are NOT sufficient.

18. a_1, a_2, a_3, ... , a_{15}

In the sequence shown, $a_n = a_{n-1} + k$, where $2 \leqslant n \leqslant 15$ and k is a nonzero constant. How many of the terms in the sequence are greater than 10 ?

(1) $a_1 = 24$

(2) $a_8 = 10$

(A) Statement (1) ALONE is sufficient, but statement (2) alone is not sufficient.

(B) Statement (2) ALONE is sufficient, but statement (1) alone is not sufficient.

(C) BOTH statements TOGETHER are sufficient, but NEITHER statement ALONE is sufficient.

(D) EACH statement ALONE is sufficient.

(E) Statements (1) and (2) TOGETHER are NOT sufficient.

$19_{\blacksquare}$ Is $m + z > 0$?

   (1) $m - 3z > 0$

   (2) $4z - m > 0$

   (A) Statement (1) ALONE is sufficient, but statement (2) alone is not sufficient.

   (B) Statement (2) ALONE is sufficient, but statement (1) alone is not sufficient.

   (C) BOTH statements TOGETHER are sufficient, but NEITHER statement ALONE is sufficient.

   (D) EACH statement ALONE is sufficient.

   (E) Statements (1) and (2) TOGETHER are NOT sufficient.

$20_{\blacksquare}$ If $xy \neq 0$ and $x^2 y^2 - xy = 6$, which of the following could be y in terms of x?

   I. $\dfrac{1}{2x}$ II. $-\dfrac{2}{x}$ III. $\dfrac{3}{x}$

   (A) I only   (B) II only   (C) I and II

   (D) I and III   (E) II and III

代数练习答案及解析

1. C。

根据 $a_n = a_{n-1} + 5$；$a_5 = a_4 + 5 = a_3 + 5 + 5 = a_2 + 5 + 5 + 5 = a_1 + 5 + 5 + 5 + 5 = a_1 + 20 = 31$，求得 $a_1 = 11$。

2. E。

解这种问题简单直接不易出错的方法是列一元一次方程。假设容器的容量为 L，根据题意：$\frac{4}{5} \times L = 4000 + \frac{3}{4} \times L$，解方程得出 $L = 80,000$。

3. C。

等差数列求和，从第 2 排开始到第 17 排停车位形成了一个公差为 2 的等差数列，$S = n(a_1 + a_n) = a_1 \times n + \frac{n \times (n-1) \times d}{2}$；从第 2 排到第 17 排的排数和为：$n = 17 - 2 + 1 = 16$，$d = 2$，$a_1 = 21$；$S = 16 \times 21 + \frac{1}{2} \times 16 \times 15 \times 2 = 576$；再加上第一排的停车位，$576 + 20 = 596$。

4. C。

等比数列 a_n，$a_n = 0.5 \times a_{n-1}$；$a_{10} \in (0.0001, 0.001)$；$a_{12} = 0.5 \times a_{11} = 0.5 \times 0.5 \times a_{10} = 0.25 \times a_{10}$。所以 $a_{10} = 4a_{12} \in (0.0001, 0.001)$；$a_{12} \in \left(\frac{0.0001}{4}, \frac{0.001}{4}\right)$，即 $a_{12} \in (0.000025, 0.00025)$。

5. A。

35 到 85 的奇数，构成一个等差数列，公差是 2。奇数的个数为：$\frac{85-35}{2} + 1 = 26$。等差数列的和为：$S = \frac{n \times (a_1 + a_n)}{2} = \frac{26 \times (35 + 85)}{2} = 1560$。

6. D。

$98 \ominus 33 = 32$，$32 \ominus 17 = 15$；$33 \ominus 17 = 16$，$98 \ominus 16 = 2$；原式 $= 15 - 2 = 13$。

7. C。

$440 = 2 \times 2 \times 2 \times 5 \times 11$，the prime sum $= 2 + 2 + 2 + 5 + 11 = 22$，不大于 35；$512 = 2^9$，the prime sum $= 2 \times 9 = 18$，不大于 35；$620 = 2 \times 2 \times 5 \times 31$，the prime sum $= 2 + 2 + 5 + 31 = 40$，大于 35；$700 = 2 \times 2 \times 5 \times 5 \times 7$，the prime sum $= 2 + 2 + 5 + 5 + 7 = 21$，不大于 35；$750 = 2 \times 3 \times 5 \times 5 \times 5$，the prime sum $= 2 + 3 + 5 + 5 + 5 = 20$，不大于 35。

8. D。

$x^2 - 2x - 15 = 0$ 解得 $x = 5$ 或者 $x = -3$，又因为 $x > 0$，所以 $x = 5$；代入三个代数式，第二个和第三个为 0。

9. A。

本题是求 $(x + y)^2$ 的值，展开得 $x^2 + 2xy + y^2$。

条件 1，已知 $xy = 10$ 和 $x^2 + y^2 = 29$，所以 $x^2 + 2xy + y^2 = 10 \times 2 + 29 = 49$，故条件 1 充分。

条件 2，仅知道 $x = 5$，不能确定 y 的值，无法计算，故条件 2 不充分。

10. D。

条件 1，$x - 3$ 是 $x^2 + bx + 12$ 的一个因数，又因为 x^2 的系数为 1，所以，设另一个因数为 $(x + k)$，则有 $(x - 3)(x + k) = 0$，展开得 $x^2 + (k - 3)x - 3k = 0$，$-3k = 12$，解得 $k = -4$，所以 $b = k - 3 = -7$。题目可求解，故条件 1 充分。

条件 2，将 $x = 4$ 代入 $x^2 + bx + 12 = 0$，解得 $b = -7$。题目可求解，故条件 2 充分。

11. B。

要判断 $\sqrt{n} > 100$，即判断是否满足 $n > 100^2$。

条件 1，$\sqrt{n - 1} > 99$，两边同时平方得 $n - 1 > 99^2$，即 $n > (99 + 1)^2 - 99 \times 2$，即 $n > 100^2 - 198$，不满足 $n > 100^2$，故条件 1 不充分。

条件 2，$\sqrt{n + 1} > 101$，两边同时平方得 $n + 1 > 101^2$，即 $n > (101 + 1)(101 - 1)$，满足 $n > 100^2$，故条件 2 充分。

185

12 ▪ E。

本题是求序列 S 前 4 项的和。

条件 1，前两项之后，每一项为其前两项的平均值，但由于前两项的值不知道，所以无法求得前 4 项的和，故条件 1 不充分。

条件 2，给出了前 3 项的平均值是 10，所以前三项的和为 30，但不知道第 4 项的值，也就无法求前 4 项的和，故条件 2 不充分。

综合条件 1 和条件 2，由于不知道前两项的值，也无法计算第 4 项的值，前 4 项的和也无法计算故条件 1 + 条件 2 不充分。

13 ▪ C。

题目要求得供应函数和需求函数相交的点，因此，联立两个函数，$m_1 x + b_1 = m_2 x + b_2$，解得 $x = \dfrac{b_2 - b_1}{m_1 - m_2}$，条件 1 只提供了 m_1 和 m_2 的值，条件 2 只提供了 $b_2 - b_1$ 的值，单独均不能解题，故均不充分。结合条件 1 和条件 2，$x = \dfrac{6}{0.005 + 0.005} = 600$，故条件 1 + 条件 2 充分。

14 ▪ D。

$x^* = ax + a$，由于未知量 a 的存在，需要求得 a 才能算出 x^*，所以 $2^* = 2a + a = 3a$。

条件 1，$3^* = 3a + a = 4a = 2$，$a = 0.5$，所以 $2^* = 3a = 1.5$。题目可求解，故条件 1 充分。

条件 2，$5^* = 5a + a = 6a = 3$，$a = 0.5$，所以 $2^* = 3a = 1.5$。题目可求解，故条件 2 充分。

15 ▪ D。

$[y]$ 表示小于或等于 y 的最大整数。

条件 1，$[y]$ 是不超过 y 的最大整数，所以 $y - [y]$ 的最小值为 0。$y - [y]$ 最大值必然小于 1，故条件 1 充分。

条件2，$[d]=0$，即不超过 d 的最大整数为 0，所以 d 的取值范围为 $0<d<1$，因为如果 $2>d\geq 1$，则 $[d]=1$，所以 $[d]=0$ 满足要求，故条件 2 充分。

16 ▪⊪ A。

前 k 项的和为：$\left(\dfrac{1}{1}-\dfrac{1}{2}\right)+\left(\dfrac{1}{2}-\dfrac{1}{3}\right)+\left(\dfrac{1}{3}-\dfrac{1}{4}\right)+\cdots+\left(\dfrac{1}{k}-\dfrac{1}{k+1}\right)=1-\dfrac{1}{k+1}=$

$\dfrac{k}{k+1}$，求和过程中中间项的前半部分和前一项的后半部分相互抵消。要使得前 k 项

的和大于 $\dfrac{9}{10}$，则 $\dfrac{k}{k+1}>\dfrac{9}{10}$，解得 $k>9$。

条件 1 给出 $k>10$，满足 $k>9$ 的条件，能够判断前 k 项的和大于 $\dfrac{9}{10}$，题目可求解，

故条件 1 充分。

条件 2 给出 $k<19$，但不能保证 $k>9$，因此也不能判断前 k 项的和是否大于 $\dfrac{9}{10}$。即

若 $9<k<19$，则前 k 项的和大于 $\dfrac{9}{10}$，否则前 k 项的和小于 $\dfrac{9}{10}$。故条件 2 不充分。

17 ▪⊪ A。

序列中除前两项为初始项，其后的每一项都为其前面
相邻两项的和，那么序列前 7 项如表所示。
条件 1，第 6 项减第 4 项等于 5，又由于第 4 项加第 5
项等于第 6 项，所以第 5 项就是 5，故条件 1 充分。

条件 2，第 6 项加第 7 项等于 21，即 $3a+5b+5a+$
$8b=8a+13b=21$，由于不知道 a 和 b 的值，也无法确
定第 5 项的值，故条件 2 不充分。

n	nth term of sequence S
1	a
2	b
3	$a+b$
4	$a+2b$
5	$2a+3b$
6	$3a+5b$
7	$5a+8b$

18 ▪⊪ B。

条件 1，知道首项为 24，但不知道公差，如果公差大于零，那么 15 个数都大于零，
如果公差小于零，则无法确定数列中小于 10 的具体个数。故条件 1 不充分。
条件 2，数列的中项 $a_8=10$，前后均有 7 个数，无论公差 k 大于零还是小于零，中
项两边总有一边是恒大于 10 的，所以一定有 7 项大于 10。故条件 2 充分。

19 ■⊪ C。

条件 1，$m-3z>0$ 可以变形为：$m+z-z-3z>0$，即 $m+z>4z$，由于不知道 z 的正负情况，所以不能确定 $m+z$ 是否大于 0，故条件 1 不充分。

条件 2，$4z-m>0$ 可以变形为：$4z-(m+z)+z>0$，即 $m+z<5z$，由于不知道 z 的正负情况，同样无法确定 $m+z$ 是否大于 0，故条件 2 不充分。

条件 1＋条件 2，首先将 $m-3z>0$ 和 $4z-m>0$ 两个式子相加，得到 $z>0$，由上面的两个式子得到 $m+z>4z$，从而得出 $m+z>0$，故条件 1＋条件 2 充分。

20 ■⊪ E。

将 xy 先看作一个整体即可算得。

第四章

几 何

4.1 ▸ 直线、角、垂线、平行线、凸多边形

直线

在几何中,直线是一条向两端无限延伸的线。

上图中,l 是直线(line);PQ 叫作线段(line segment);P 和 Q 两点叫作端点(end point)。

角

如果两条直线相交,会形成两组对顶角(vertical angles),对顶角本身是相等的。

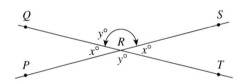

上图中 $\angle QRP$ 和 $\angle SRT$ 是对顶角;$\angle QRS$ 和 $\angle PRT$ 是对顶角。因为 PS 是直线,所以 $x + y = 180$。

垂线

当两条直线相交所形成的夹角为90°(right angle)时,两条直线是互相垂直的。例如:

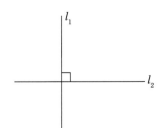

平行线

如果两条直线在同平面上且永不相交,则这两条直线是平行线。

——————————————————————— l_1

——————————————————————— l_2

如果这两条直线被第三条直线穿过,则会形成同位角、内错角和同旁内角这三个概念。

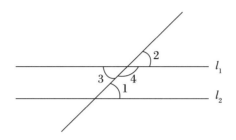

如图所示,∠1 和∠2 是同位角,两者相等;∠1 和∠3 是内错角,两者也相等;∠1 和∠4 是同旁内角,两者相加等于 $180°$。

凸多边形

凸多边形的每个内角都是小于$180°$的。凸多边形只需记住内角和公式即可:

$$n \text{ 边形内角和} = (n-2) \times 180°。$$

例题 1

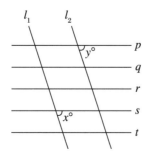

If $l_1 /\!/ l_2$ in the figure above, is $x = y$?

(1) $p /\!/ r \ and \ r /\!/ t$

(2) $q /\!/ s$

（A）Statement（1）ALONE is sufficient, but statement（2）alone is not sufficient.

（B）Statement（2）ALONE is sufficient, but statement（1）alone is not sufficient.

（C）BOTH statements TOGETHER are sufficient, but NEITHER statement ALONE is sufficient.

（D）EACH statement ALONE is sufficient.

（E）Statements（1）and（2）TOGETHER are NOT sufficient.

解：

想知道 x 和 y 是否相等，我们需要先知道 p，q，r，s 这四条直线是否平行。如果它们都平行，则 x 必然等于 y；如果它们之间不平行，则无法确定 x 和 y 是否相等。

条件 1 说，p，r，t 三者平行。由于不知道直线 s 的情况，所以条件 1 不充分。

条件 2 说，q 和 s 平行。由于不知道 p 的情况，所以条件 2 不充分。

两个条件同时成立时，我们依然不知道 p，q，r，s 这四条直线是否平行，所以条件 1 + 条件 2 也不充分。

综上，答案为 E。

例题 2

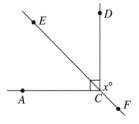

In the figure above, if F is a point on the line that bisects angle ACD and the measure of angle DCF is $x°$, which of the following is true of x?

（A）$90 < x < 100$ （B）$100 < x < 110$ （C）$110 < x < 120$

（D）$120 < x < 130$ （E）$130 < x < 140$

解：

F 点所在的直线是 $\angle ACD$ 的平分线，平分的两个角分别是 $45°$。$\angle ECD = 45°$，因为 ECF 是一条直线 $180°$，所以，$\angle x = 180° - 45° = 135°$。

因此，$130 < x < 140$。

综上，答案是 E。

例题 3

In pentagon $PQRST$, $PQ = 3$, $QR = 2$, $RS = 4$, and $ST = 5$. Which of the lengths 5, 10, and 15 could be the value of PT?

（A） 5 only   （B） 15 only   （C） 5 and 10 only

（D） 10 and 15 only   （E） 5，10，and 15

解：

因为多边形各边的夹角必须小于 $180°$，所以任何一边必然小于另外所有边的和（你可以假想把各条边都拉平，这是一条边的极限长度）。因此，本题中的五边形的最后一条边必然小于 $3 + 2 + 4 + 5 = 14$。即 PT 的长度必须小于 14，故只能选 5 和 10，答案为 C。

4.2 ▸ 三角形

三边构成的多边形被称为三角形。在三角形中，任意两边之和大于第三边；任意两边之差小于第三边。

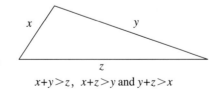

$$x+y>z,\ x+z>y \text{ and } y+z>x$$

三角形是多边形中的一种，越小的角对应的边也越小，反之亦然。

在三角形中，等腰三角形、等边三角形和直角三角形为三种特殊三角形。所谓等腰三角形指的是两个腰相等、两个底角也相等的三角形。所谓等边三角形指的是三角形的三边都相等、每个内角都是 $60°$ 的三角形。

193

经常考到的是直角三角形。所有的直角三角形都满足"勾股定理"。直角三角形的两个直角边边长的平方之和等于斜边长的平方。如果设直角三角形的两条直角边长度分别是 a 和 b，斜边长度是 c，那么可以用数学语言表达为：

$$a^2 + b^2 = c^2。$$

最经典的两组直角三角形的三边为：3，4，5 和 6，8，10。

三角函数在题目中也偶有涉及，但是较之高中的考查难度可是低多了，一般只有对特殊角的数值考查。另外需记住，sin 是"对边比斜边"；cos 是"邻边比斜边"；tan 是"对边比邻边"。

三角函数值 三角函数 角	0°	30°	45°	60°	90°
$\sin\alpha$	0	$\dfrac{1}{2}$	$\dfrac{\sqrt{2}}{2}$	$\dfrac{\sqrt{3}}{2}$	1
$\cos\alpha$	1	$\dfrac{\sqrt{3}}{2}$	$\dfrac{\sqrt{2}}{2}$	$\dfrac{1}{2}$	0
$\tan\alpha$	0	$\dfrac{\sqrt{3}}{3}$	1	$\sqrt{3}$	不存在
$\cot\alpha$	不存在	$\sqrt{3}$	1	$\dfrac{\sqrt{3}}{2}$	0

三角形的面积公式为：$\dfrac{底 \times 高}{2}$。

例题 1

In isosceles triangle PQR, if the measure of angle P is $80°$, which of the following could be the measure of angle R?

Ⅰ. $20°$ Ⅱ. $50°$ Ⅲ. $80°$

（A）Ⅰ only   （B）Ⅲ only   （C）Ⅰ and Ⅱ only

（D）Ⅱ and Ⅲ only   （E）Ⅰ，Ⅱ，and Ⅲ

解：

等腰 $\triangle PQR$ 的某个角为 $80°$。如果这个角为底角，$\angle PRQ$ 也可以为底角，同为 $80°$；$\angle PRQ$ 也可以为顶角，即 $20°$；如果这个 $80°$ 的角为顶角，则两个底角均为 $50°$，$\angle PRQ$ 可以为底角之一。因此，答案为 E。

例题 2

If two sides of a triangle have lengths 2 and 5, which of the following could be the perimeter of the triangle?

Ⅰ. 9　　　Ⅱ. 15　　　Ⅲ. 19

（A）None　　　　　　（B）Ⅰ only　　　　　　（C）Ⅱ only

（D）Ⅱ and Ⅲ only　　（E）Ⅰ, Ⅱ, and Ⅲ

解：

三角形两边之和需大于第三边。因为 $2+5=7$，不大于任何一个可选数据，所以本题中的三个数据均不可能是该三角形的第三边，故答案为 A。

例题 3

A certain right triangle has sides of length x, y, and z, where $x < y < z$. If the area of this triangular region is 1, which of the following indicates all of the possible values of y?

（A）$y > \sqrt{2}$　　　　　（B）$\dfrac{\sqrt{3}}{2} < y < \sqrt{2}$　　　　　（C）$\dfrac{\sqrt{2}}{3} < y < \dfrac{\sqrt{3}}{2}$

（D）$\dfrac{\sqrt{3}}{4} < y < \dfrac{\sqrt{2}}{3}$　　　（E）$y < \dfrac{\sqrt{3}}{4}$

解：

本题有些难度。题目只告诉了我们直角三角形的面积为 1，根据 $x < y < z$ 可知，x 和 y 分别是两个直角边。所以 $x \times y = 2$。因为 z 的长度不限，我们可以让 x 无限小，y 无限大，只要两者乘积为 2 即可。至此，我们可以确定 y 可以无限大，也就是没有上限，这时就已经可以确定答案为 A 了。

那么最小值怎么算呢？显然，当 x 和 y 的大小尽量接近时，y 最小。等腰直角三角形的两个直角边相等，即 $y \times y = 2$。由此可知，y 的最小值为 $\sqrt{2}$。综上，答案为 A。

例题 4

In the figure shown, what is the area of triangular region *BCD*?

(A) $4\sqrt{2}$　　(B) 8　　(C) $8\sqrt{2}$　　(D) 16　　(E) $16\sqrt{2}$

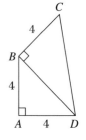

解：

△*ABD* 是等腰直角三角形，根据勾股定理得出斜边 $BD = \sqrt{4^2 + 4^2} = \sqrt{32} = 4\sqrt{2}$，

$$S_{BCD} = \frac{1}{2} \times 4 \times 4\sqrt{2} = 8\sqrt{2}。$$

综上，答案为 C。

例题 5

In the figure shown, *PQRSTU* is a regular polygon with sides of length *x*. What is the perimeter of triangle *PRT* in terms of *x*?

(A) $\dfrac{x\sqrt{3}}{2}$　　(B) $x\sqrt{3}$　　(C) $\dfrac{3x\sqrt{3}}{2}$

(D) $3x\sqrt{3}$　　(E) $4 \times \sqrt{3}$

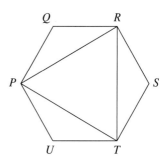

解：

依题意，由于 *PQRSTU* 是正六边形，所以三角形 *PRT* 必然为等边三角形。三角形 *PQR*，*RST* 和 *PUT* 为全等三角形。六边形的内角和为 $4 \times 180° = 720°$，则正六边形每个角为 $120°$。又因为 $\angle RPT$ 为 $60°$，所以 $\angle QPR = \angle TPU = 30°$。

$$\cos 30° = \frac{\dfrac{PR}{2}}{x}$$

因此，$PR = x \times \sqrt{3}$；三角形 *PRT* 的周长为 $3 \times x \times \sqrt{3}$。

综上，答案为 D。

例题 6

If the lengths of the legs of a right triangle are integers, what is the area of the triangular region?

(1) The length of one leg is $\dfrac{3}{4}$ the length of the other.

(2) The length of the hypotenuse is 5.

(A) Statement (1) ALONE is sufficient, but statement (2) alone is not sufficient.

(B) Statement (2) ALONE is sufficient, but statement (1) alone is not sufficient.

(C) BOTH statements TOGETHER are sufficient, but NEITHER statement ALONE is sufficient.

(D) EACH statement ALONE is sufficient.

(E) Statements (1) and (2) TOGETHER are NOT sufficient.

解:

条件 1 说，直角三角形的两个直角边之比为 3:4。显然，当三边为 3，4，5 和 6，8，10 的时候，直角边之比均为 3:4，但面积显然不同。故条件 1 不充分。

条件 2 说，斜边长度为 5。直角三角形斜边长度为 5，则另外两边必为 3 和 4，面积确定。故条件 2 充分。

综上，答案为 B。

例题 7

If x and y are the lengths of the legs of a right triangle, what is the value of xy?

(1) The hypotenuse of the triangle is $10\sqrt{2}$.

(2) The area of the triangular region is 50.

(A) Statement (1) ALONE is sufficient, but statement (2) alone is not sufficient.

(B) Statement (2) ALONE is sufficient, but statement (1) alone is not sufficient.

(C) BOTH statements TOGETHER are sufficient, but NEITHER statement ALONE is sufficient.

(D) EACH statement ALONE is sufficient.

（E） Statements （1） and （2） TOGETHER are NOT sufficient.

解：

x 和 y 是直角边。

条件 1 说，三角形的斜边是 $10\sqrt{2}$。通过斜边的值，我们可以确定 $x^2 + y^2 = 200$，但无法确定 xy 的值，故条件 1 不充分。

条件 2 说，三角形面积是 50。显然，直角三角形的面积等于两个直角边之积除以 2，由此可确定 xy 的值，故条件 2 充分。

综上，答案为 B。

例题 8

In the figure shown, *RST* is a triangle with angle measures as shown and *PRTQ* is a line segment. What is the value of $x + y$?

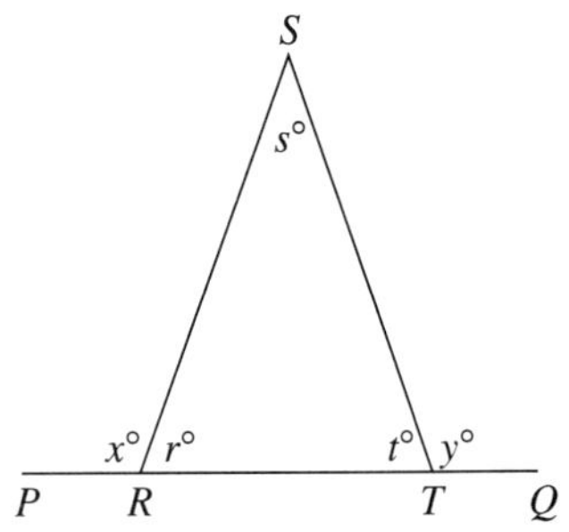

（1） $s = 40$

（2） $r = 70$

（A） Statement （1） ALONE is sufficient, but statement （2） alone is not sufficient.

（B） Statement （2） ALONE is sufficient, but statement （1） alone is not sufficient.

（C） BOTH statements TOGETHER are sufficient, but NEITHER statement ALONE is sufficient.

（D） EACH statement ALONE is sufficient.

（E） Statements （1） and （2） TOGETHER are NOT sufficient.

解：

条件 1，s 是 $40°$，由于三角形的内角和为 $180°$，所以 $180 - s = r + t$。又因为 *PRTQ* 是直线，所以 $x + y = 180 - r + 180 - t$。由此可确定 $x + y$ 的值，故条件 1 充分。

条件 2，只知道 r 的值，无法得知 t 的值，故条件 2 不充分。

综上，答案为 A。

例题 9

In the triangle shown, is $x > 90$?

(1) $a^2 + b^2 < 15$

(2) $c > 4$

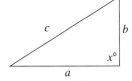

(A) Statement (1) ALONE is sufficient, but statement (2) alone is not sufficient.

(B) Statement (2) ALONE is sufficient, but statement (1) alone is not sufficient.

(C) BOTH statements TOGETHER are sufficient, but NEITHER statement ALONE is sufficient.

(D) EACH statement ALONE is sufficient.

(E) Statements (1) and (2) TOGETHER are NOT sufficient.

解：

题目问的是 x 是否大于 90。

条件 1 说，两条边长度的平方和小于 15，由于不知道斜边的长度是多少，所以无法确定 x 的值，故条件 1 不充分。

条件 2 说，c 大于 4。只知道斜边的长度，无法确定角度，故条件 2 不充分。

两个条件同时成立时，如果 c 大于 4，则 c^2 必定大于 16。结合条件 1，可以知道 $c^2 > a^2 + b^2$。若 $a^2 + b^2 = c^2$，则 $x = 90$。若 $c^2 > a^2 + b^2$，则 x 必定大于 90。故条件 1 + 条件 2 充分。

请注意，在平面图形中，必然是大角对大边，即角度大小和边长大小成正比。

综上，答案为 C。

例题 10

In triangle ABC shown, what is the length of side BC?

(1) Line segment AD has length 6.

(2) $x = 36$

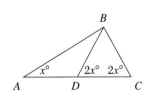

（A）Statement（1）ALONE is sufficient, but statement（2）alone is not sufficient.

（B）Statement（2）ALONE is sufficient, but statement（1）alone is not sufficient.

（C）BOTH statements TOGETHER are sufficient, but NEITHER statement ALONE is sufficient.

（D）EACH statement ALONE is sufficient.

（E）Statements（1）and（2）TOGETHER are NOT sufficient.

解：

依题意，$\angle ABD = \angle A = x$。由此可知，$AD = BD = BC$。

条件1说，AD 的长度为6。显然可以知道 BC 的长度，故条件1充分。

条件2说，$x = 36$。这个条件只给出了角度，没有给出边长，故条件2不充分。

综上，答案为A。

例题 11

The perimeter of a certain isosceles right triangle is $16 + 16\sqrt{2}$. What is the length of the hypotenuse of the triangle?

（A）8　　　（B）16　　　（C）$4\sqrt{2}$　　　（D）$8\sqrt{2}$　　　（E）$16\sqrt{2}$

解：

本题最重要的是知道 isosceles right triangle 的意思。这个词是"等腰直角三角形"的意思。题目问的是斜边长度。

等腰直角三角形两条直角边和斜边的比是 $1:1:\sqrt{2}$（勾股定理），周长为 $16 + 16\sqrt{2}$，设直角边长为 a，那么 $2a + \sqrt{2}a = 16 + 16\sqrt{2}$，求解得 $a = 8\sqrt{2}$，那么斜边长为 $\sqrt{2}a = 16$。

答案为B。

例题 12

If each side of $\triangle ACD$ above has length 3 and if AB has length 1, what is the area of region $BCDE$?

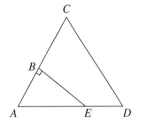

(A) $\dfrac{9}{4}$　　　(B) $\dfrac{7}{4}\sqrt{3}$　　　(C) $\dfrac{9}{4}\sqrt{3}$

(D) $\dfrac{7}{2}\sqrt{3}$　　　(E) $6+\sqrt{2}$

解：

$\triangle ACD$ 是等边三角形，因此 $\angle A$ 为 $60°$。根据三角函数定义可知，

$$\tan a = \frac{BE}{AB} = BE = \sqrt{3},$$

$\triangle ABE$ 的面积为 $\dfrac{\sqrt{3}}{2}$。

$\triangle ACD$ 的面积可以用等边三角形的边长和面积关系求得，

如果 a 是等边三角形的边长，那么它的面积 $= \dfrac{1}{2} \times \dfrac{\sqrt{3}a}{2} \times a = \dfrac{\sqrt{3}a^2}{4}$。

因此，$\triangle ACD$ 的面积为 $\dfrac{9\sqrt{3}}{4}$。

四边形 $BCDE$ 的面积为两者相减，即

$$\frac{9\sqrt{3}}{4} - \frac{\sqrt{3}}{2} = \frac{7\sqrt{3}}{4}。$$

答案为 B。

例题 13

In the figure shown, what is the perimeter of $\triangle PQR$?

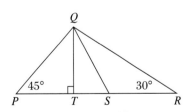

(1) The length of segment PT is 2.

(2) The length of segment RS is $\sqrt{3}$.

(A) Statement (1) ALONE is sufficient, but statement (2) alone is not sufficient.

(B) Statement (2) ALONE is sufficient, but statement (1) alone is not sufficient.

(C) BOTH statements TOGETHER are sufficient, but NEITHER statement ALONE is sufficient.

(D) EACH statement ALONE is sufficient.

(E) Statements (1) and (2) TOGETHER are NOT sufficient.

解:

题目问的是 $\triangle PQR$ 的周长。

条件1说,PT 的长度是2。通过 PT 的长度和 $\angle P$ 的值,可以求出 QT 和 QP 的长度。由 QT 的长度和 $\angle R$ 的值,又可以求出 RT 和 QR 的长度。因此,$\triangle PQR$ 的三边都已知,周长可求,故条件1充分。

条件2说,RS 的长度是 $\sqrt{3}$。我们无法知道 RS 的长度是否和 QS 的长度相等,所以无法推断 QT 和 QR 的长度,故条件2不充分。

综上,答案为 A。

例题 14

Positions of the same ladder leaning against the side SV of a wall. The length of TV is how much greater than the length of RV?

(1) The length of TU is 10 meters.

(2) The length of RV is 5 meters.

(A) Statement (1) ALONE is sufficient, but statement (2) alone is not sufficient.

(B) Statement (2) ALONE is sufficient, but statement (1) alone is not sufficient.

(C) BOTH statements TOGETHER are sufficient, but NEITHER statement ALONE is sufficient.

(D) EACH statement ALONE is sufficient.

(E) Statements (1) and (2) TOGETHER are NOT sufficient.

解：

题目问的是 TV 比 RV 长多少，也就是问 TR 的长度。请注意，题干中明确表示，TU 和 RS 是相同的梯子。也就是说，$\triangle UTV$ 和 $\triangle SRV$ 的斜边长度相等。

条件 1 说，TU 的长度为 10，这表示 SR 的长度也为 10。知道直角三角形的一条边和角度，肯定可以通过三角函数求出 TV 和 RV 的值，故条件 1 充分。

条件 2 说，RV 等于 5。通过 RV 和 $\angle SRV$ 可以求出 SR 的长度，也就是可以求出 TU 的长度。根据 $\angle UTV$ 和 TU 的长度，必然可以求出 TV 的长度，故条件 2 充分。

综上，答案为 D。

例题 15

What is the length of the hypotenuse of $\triangle ABC$?

(1) The lengths of the three sides of $\triangle ABC$ are consecutive even integers.

(2) The hypotenuse of $\triangle ABC$ is 4 units longer than the shorter leg.

(A) Statement (1) ALONE is sufficient, but statement (2) alone is not sufficient.

(B) Statement (2) ALONE is sufficient, but statement (1) alone is not sufficient.

(C) BOTH statements TOGETHER are sufficient, but NEITHER statement ALONE is sufficient.

(D) EACH statement ALONE is sufficient.

(E) Statements (1) and (2) TOGETHER are NOT sufficient.

解：

题目问的是直角三角形的斜边长度。hypotenuse 特指直角三角形的斜边。

条件 1 说，$\triangle ABC$ 的三边长度是连续的偶数。三边长度是连续偶数的直角三角形，只有 6，8，10 这一种情况，因此斜边长度必为 10，故条件 1 充分。

条件 2 说，斜边比更短的直角边长 4 个单位。我们只能根据勾股定理列出关系式（设 x 为短直角边，y 为长直角边）：

$$x^2 + y^2 = (x+4)^2。$$

显然无法确定 x 和 y 的值，故条件 2 不充分。

答案为 A。

4.3 ▶ 四边形

顾名思义，四边形就是有四条边的图形。四边形的考有内容非常单一，只考查几种特殊四边形的性质。

平行四边形

平行四边形的对边相等且平行，形如：

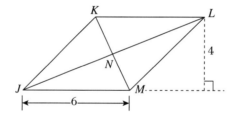

平行四边形的对角线（diagonal）被互相平分，即 $KN = NM$；$JN = NL$。

平行四边形的面积（area）等于底乘高，即 $6 \times 4 = 24$。

矩形

矩形是特殊的平行四边形。如果一个平行四边形的每个内角都是 90°，那么它就是矩形。

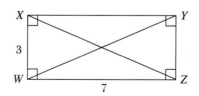

如果矩形每条边的长度相等，那么它就是正方形（square）。

梯形

四边形中只有两条边平行的叫梯形，面积为"上底和下底的和乘高除以2"。

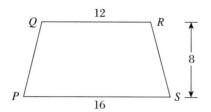

上图梯形的面积为：$\dfrac{(12+16)\times 8}{2}=112$。

菱形

四边相等的四边形是菱形，正方形是特殊的菱形。菱形的对角线相互垂直。

有一条特殊的性质需要注意：

多边形对角线的数量 $=\dfrac{n(n-3)}{2}$（n 是边的数量）。

例题 1

If the length of a diagonal of a square is $2\sqrt{x}$, what is the area of the square in terms of x?

（A）$\sqrt{x}$　　　（B）$\sqrt{2}x$　　　（C）$2\sqrt{x}$　　　（D）x　　　（E）$2x$

解：

对角线和正方形的两条边会构成直角三角形，设正方形的边长为 k，则有：

$$2k^2=4x,$$

解得

$$k^2=2x,$$

也就是说，正方形的面积为 $2x$。

答案为 E。

例题 2

In the figure shown, line segment QR has length 12, and rectangle $MPQT$ is a square. If the area of rectangular region $MPRS$ is 540, what is the area of rectangular region $TQRS$?

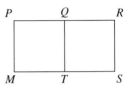

Note: Not drawn to scale.

(A) 144   (B) 216   (C) 324

(D) 360   (E) 396

解:

$MPQT$ 是正方形,假设边长为 a,则正方形 $MPQT$ 的面积是 a^2。

QT 是 $MPQT$ 和 $QTSR$ 的共同边,因此 $QTSR$ 的面积是 $a \times 12$。

这两个四边的面积之和等于整个大长方形 $PMRS$ 的面积,即 $a^2 + 12 \times a = 540$,

可以求得 $a = 18$。

所以 $TQRS$ 的面积为 $18 \times 12 = 216$,答案为 B。

例题 3

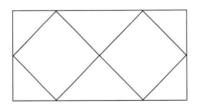

In the figure shown, two identical squares are inscribed in the rectangle. If the perimeter of the rectangle is $18\sqrt{2}$, then what is the perimeter of each square?

(A) $8\sqrt{2}$ (B) 12   (C) $12\sqrt{2}$ (D) 16   (E) 18

解:

本题具有一定的难度,需要作两条辅助线。因为我们只知道外围大长方形的周长,所以需要用内接的小正方形的边长来分别表示大长方形的长与宽。辅助线如下图:

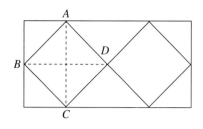

设正方形的边长为 a，则对角线 AC 和 BD 的长度均为 $\sqrt{2}a$；AC 等于外围大长方形的宽；BD 是外围大长方形的长的 $\frac{1}{2}$，因此大长方形的长为 $2\sqrt{2}a$。由此可知，$\sqrt{2}a + 2\sqrt{2}a = 9\sqrt{2}$，$a = 3$。整个正方形的周长为 12，答案为 B。

例题 4

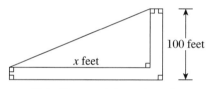

Note: Figure not drawn to scale.

The figure above shows some of the dimensions of a triangular plaza with an L-shaped walk along two of its edges. If the width of the walk is 4 feet and the total area of the plaza and walk together is 10,800 square feet, what is the value of x?

(A) 200　　　(B) 204　　　(C) 212　　　(D) 216　　　(E) 225

解：

由于步道宽为 4，所以三角形的直角边分别为 x 和 96，步道由两个长方形组成。总面积为：

$$\frac{96x}{2} + 100 \times 4 + 4x = 10800,$$

$$x = 200。$$

答案为 A。

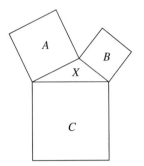

例题 5

In the figure shown, three squares and a triangle have areas of A, B, C, and X as shown. If $A = 144$, $B = 81$ and $C = 225$, then $X =$

(A) 150　　　(B) 144　　　(C) 80

(D) 54　　　(E) 36

Note: Figure not drawn to scale.

解：

已知条件给出了 ABC 三个正方形的面积，由此可知三角形 X 的三边为 9，12，15。

三边比为 3:4:5 时，X 必为直角三角形。因此，X 面积为 54。故答案为 D。

例题 6

In the figure shown, the area of the shaded region is

(A) $8\sqrt{2}$　　　　(B) $4\sqrt{3}$　　　　(C) $4\sqrt{2}$

(D) $8(\sqrt{3}-1)$　　　(E) $8(\sqrt{2}-1)$

解：

这道题目的解题关键在于准确画出辅助线。

连接点 A 和点 B。依题意，四边形 $ACBD$ 是正方形，线段 AB 是对角线。

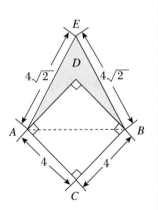

由于正方形的边长为 4，所以对角线 AB 的长度为 $4\sqrt{2}$。也就是说，三角形 EAB 是边长为 $4\sqrt{2}$ 的等边三角形。阴影部分的面积是该三角形的面积减去正方形面积的一半。

如果 a 是等边三角形的边长，那么它的面积 $= \dfrac{\sqrt{3}a \times a}{4}$。

阴影部分面积 $= \dfrac{\sqrt{3}a \times a}{4} - 8 = 8(\sqrt{3}-1)$。

答案为 D。

例题 7

The shaded region in the figure shown represents a rectangular frame with length 18 inches and width 15 inches. The frame encloses a rectangular picture that has the same area as the frame itself. If the length and width of the picture have the same ratio as the length and width of the frame, what is the length of the picture, in inches?

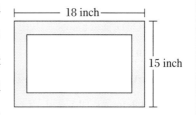

(A) $9\sqrt{2}$　　(B) $3\sqrt{2}$　　(C) $\dfrac{9}{\sqrt{2}}$　　(D) $15\left(1-\dfrac{1}{\sqrt{2}}\right)$　　(E) $\dfrac{9}{2}$

解：

依题意，设空白部分的两边长度分别为 $18k$ 和 $15k$。题干中说阴影部分和空白部分面积相等，也就是说，空白部分面积的两倍等于整体面积，即

$$18k \times 15k \times 2 = 18 \times 15,$$

解出：

$$k = \frac{\sqrt{2}}{2},$$

空白部分的长为：

$$18k = 9\sqrt{2},$$

答案为 A。

例题 8

Sprinklers are being installed to water a lawn. Each sprinkler waters in a circle. Can the lawn be watered completely by 4 installed sprinklers?

(1) The lawn is rectangular and its area is 32 square yards.

(2) Each sprinkler can completely water a circular area of lawn with a maximum radius of 2 yards.

（A）Statement（1）ALONE is sufficient, but statement（2）alone is not sufficient.

（B）Statement（2）ALONE is sufficient, but statement（1）alone is not sufficient.

（C）BOTH statements TOGETHER are sufficient, but NEITHER statement ALONE is sufficient.

（D）EACH statement ALONE is sufficient.

（E）Statements（1）and（2）TOGETHER are NOT sufficient.

解：

条件1说，长方形草场的面积为32。我们不知道喷头喷水范围的半径多大，所以不知道是否能覆盖整个长方形草场区域。故条件1不充分。

条件2说，喷头喷水范围的最大半径是2。我们知道最大半径，但不知道草场的大小，因此条件2不充分。

条件1＋条件2，按照最大半径算，喷头喷水范围可以覆盖整个草场。但若不是最大半径，则不一定覆盖。因此条件1＋条件2依然不充分。

综上，答案为E。

所谓喷头喷水范围的最大半径，就是条件2里给出的2 yards。根据条件2，我们能算出这些喷头喷水范围能覆盖的最大面积。因为这个面积超过了长方形草场的面积，所以在这个情况下是可以全部覆盖长方形草场的。但如果喷头喷水范围的最大半径比2 yards小，譬如变成0.1 yards，就不能全部覆盖长方形草场了。

例题 9

In the figure shown, if the shaded region is rectangular, what is the length of XY?

（1）The perimeter of the shaded region is 24.

（2）The measure of $\angle XYZ$ is 45°.

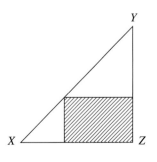

（A）Statement（1）ALONE is sufficient, but statement（2）alone is not sufficient.

（B） Statement（2）ALONE is sufficient, but statement（1）alone is not sufficient.

（C） BOTH statements TOGETHER are sufficient, but NEITHER statement ALONE is sufficient.

（D） EACH statement ALONE is sufficient.

（E） Statements（1）and（2）TOGETHER are NOT sufficient.

解：

条件 1 说，阴影区域的周长为 24。只知道阴影部分的周长，显然无法得知三角形斜边的长度，故条件 1 不充分。

条件 2 说，∠X 是 45°。只知道角度，不知道各边的长度数值，无法确定 XY 的长度，故条件 2 不充分。

两个条件同时成立时，如果阴影区域是长方形，则三角形 XYZ，XAB，AYC 均为直角三角形，如下图：

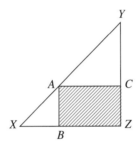

同时因为 ∠X 是 45°，所以，这三个三角形必然都是等腰直角三角形。因此，XB = AB；AC = CY。

又因为阴影部分的周长为 24，所以 XZ + ZY 等于 24。

$$XZ = ZY = 12。$$

根据勾股定理，必然可求斜边 XY 的长度，所以条件 1 和条件 2 充分，答案为 C。

例题 10

In rectangular region *PQRS* shown, *T* is a point on side *PS*. If *PS* = 4, what is the area of region *PQRS*?

（1） △*QTR* is equilateral.

（2） Segments *PT* and *TS* have equal length.

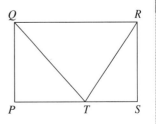

(A) Statement (1) ALONE is sufficient, but statement (2) alone is not sufficient.

(B) Statement (2) ALONE is sufficient, but statement (1) alone is not sufficient.

(C) BOTH statements TOGETHER are sufficient, but NEITHER statement ALONE is sufficient.

(D) EACH statement ALONE is sufficient.

(E) Statements (1) and (2) TOGETHER are NOT sufficient.

解：

题目问的是长方形 *PQRS* 的面积。因为已知其长为 4，所以只需要条件能告诉我们宽即可。

条件 1 说，$\triangle QTR$ 是等边三角形。由此可知，$\angle RQT = 60°$，$\angle PQT = 30°$；$QR = QT = 4$。由于已知 $\triangle QPT$ 的一个角和一条边，所以一定能通过三角函数求出 QP 的值，从而求出长方形 *PQRS* 的面积，故条件 1 充分。

条件 2 说，$PT = TS = 2$。由此我们只能判断 $QT = RT$，但无法得知 QP 和 RS 的长度。故条件 2 不充分。

综上，答案为 A。

4.4 ▶ 圆

在圆内画一条线叫作弦（chord）；如果这条弦刚好穿过了圆的圆心（center），那么它就可以叫作圆的直径（diameter）。直径的一半是半径（radius）。圆的周长（circumference）是 $2\pi r$；面积是 πr^2。

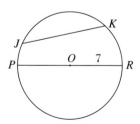

在上图中，*PR* 是直径，*OR* 是半径。如果 $OR = 7$，则周长为 $2\pi \times 7 = 14\pi$；面积为：$\pi \times 7^2 = 49\pi$。

整个圆的圆心角为 360°。

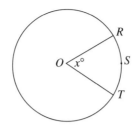

如上图，弧 RST 的长度为 $\dfrac{x}{360}$ 的圆周长。举个例子，如果圆心角 $x°$ 为 $60°$，则弧 RST 的长度为 $\dfrac{1}{6}$ 的圆周长。

如果一条直线只和圆有一个交点，那么这条直线被称为圆的切线（tangent to the circle），那个交点叫作切点（point of tangency）。从圆心到切点的连线永远垂直于切线。

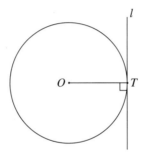

如果一个多边形的每个顶点都在圆上，则我们称这个多边形内接于圆（inscribed in the circle），形如下图。

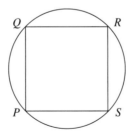

如果多边形的每一边都和圆相切，则我们称该圆内切于多边形，形如下图。

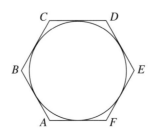

直径对应的圆周角永远是 $90°$，形如下图。

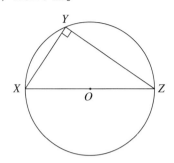

例题 1

In the figure shown, if the area of the shaded region is 3 times the area of the smaller circular region, then the circumference of the larger circle is how many times the circumference of the smaller circle?

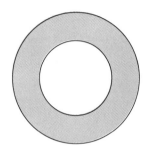

(A) 4   (B) 3   (C) 2

(D) $\sqrt{3}$ (E) $\sqrt{2}$

解：

设大圆的半径为 R，小圆的半径为 r。小圆面积为 πr^2；阴影部分的面积为 $\pi R^2 - \pi r^2$。面积比值为：$\dfrac{\pi R^2 - \pi r^2}{\pi r^2} = 3$，解得

$$R^2 = 4r^2,$$

$$R = 2r,$$

因为两个圆的半径比等于它们的周长比，故答案为 C。

例题 2

In the figure shown, the triangle is inscribed in the semicircle. If the length of line segment AB is 8 and the length of line segment BC is 6, what is the length of arc ABC?

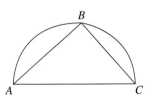

(A) 15π (B) 12π (C) 10π (D) 7π (E) 5π

解：

因为是半圆，所以 AC 必然是直径。∠B 为直角。根据勾股定理可知 AC 长度为 10。整个圆周长为 10π，则半圆的弧长为 5π，故答案为 E。

例题 3

In the figure above, O is the center of the circle. If the area of the shaded region is 2π, what is the value of x?

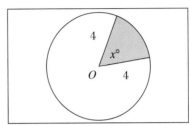

Note:Figure not drawn to scale.

(A) $\dfrac{45}{2}$　　　(B) 30　　　(C) 45

(D) 60　　　(E) 90

解：

已知半径为 4，整个圆的面积为 16π。阴影部分占整个圆面积的 $\dfrac{1}{8}$。因此，$x°$ 应该占整个圆心角的 $\dfrac{1}{8}$，即，$\dfrac{1}{8} \times 360° = 45°$，故答案为 C。

例题 4

If rectangle $ABCD$ is inscribed in the circle, what is the area of the circular region?

(A) 36.00π　　　(B) 42.25π　　　(C) 64.00π

(D) 84.50π　　　(E) 169.00π

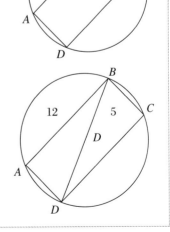

解：

本题有些难度，需要作辅助线。

连接长方形的对角线 BD，因为 ∠A 是 90°，所以 BD 必然过圆心，直线 BD 为直径。根据勾股定理可知 BD 的长度为 13，半径为 6.5。根据圆的面积公式，$\pi \times 6.5^2 = 42.25\pi$，故答案为 B。

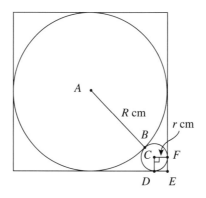

The figure above shows 2 circles. The larger circle has center A, radius R cm, and is inscribed in a square. The smaller circle has center C, radius r cm, and is tangent to the larger circle at point B and to the square at points D and F. If points A, B, C, and E are collinear, which of the following is equal to $\dfrac{R}{r}$?

(A) $\dfrac{2}{\sqrt{2}+1}$ (B) $\dfrac{2}{\sqrt{2}-1}$ (C) $\dfrac{2}{2\sqrt{2}+1}$

(D) $\dfrac{\sqrt{2}+1}{\sqrt{2}-1}$ (E) $\dfrac{2\sqrt{2}+1}{2\sqrt{2}-1}$

解：

本题的关键在于理解线段 AE 是由三个部分构成的，即 R，r 和正方形 $CDEF$ 的对角线。连接 AE，并且作点 A 到大正方形的底边垂线，垂点为 M，如下图：

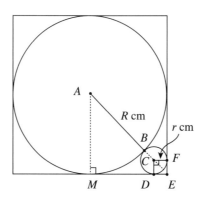

显然，$AM = R$，由于 $CDEF$ 是正方形，所以 $\angle AEM$ 等于 $45°$。因此 $AM = EM = R$。

根据勾股定理可知：

$$AE^2 = AM^2 + EM^2 = 2R^2,$$

$$AE = R + r + \sqrt{2}r = \sqrt{2}R,$$

整理该式则有：

$$\frac{R}{r} = \frac{\sqrt{2}+1}{\sqrt{2}-1}$$

故答案为 D。

In the figure shown, PQ is a diameter of circle O, $PR = SQ$, and $\triangle RST$ is equilateral. If the length of PQ is 2, what is the length of RT?

(A) $\dfrac{1}{2}$ (B) $\dfrac{1}{\sqrt{3}}$ (C) $\dfrac{2}{\sqrt{3}}$

(D) $\dfrac{2}{\sqrt{3}}$ (E) $\sqrt{3}$

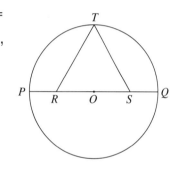

因为 $PR = SQ$，所以 $RO = OS$。连接 TO，

根据等腰（边）三角形的三线合一规则，TO 必然垂直于 RS，同时平分角 T。因为 $\triangle RST$ 是等边三角形，所以 $\angle T$ 等于 $60°$，$\angle T$ 的一半为 $30°$。

PQ 是圆的直径，所以 $PQ = 2TO = 2$，即

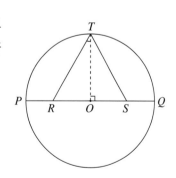

$$TO = 1,$$

$$\cos 30° = \frac{TO}{TR} = \frac{1}{TR} = \frac{\sqrt{3}}{2},$$

$$TR = \frac{2}{\sqrt{3}}。$$

答案为 D。

例题 7

In the figure shown, points A, B, C, and D are collinear, and AB, BC, and CD are semicircles with diameters d_1 cm, d_2 cm, and d_3 cm, respectively. What is the sum of the lengths of AB, BC and CD in centimeters?

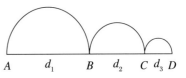

(1) $d_1:d_2:d_3$ is 3:2:1.

(2) The length of AD is 48 cm.

(A) Statement (1) ALONE is sufficient, but statement (2) alone is not sufficient.

(B) Statement (2) ALONE is sufficient, but statement (1) alone is not sufficient.

(C) BOTH statements TOGETHER are sufficient, but NEITHER statement ALONE is sufficient.

(D) EACH statement ALONE is sufficient.

(E) Statements (1) and (2) TOGETHER are NOT sufficient.

解：

条件 1 只告诉了我们三个半圆直径的比值，故条件 1 不充分。

条件 2 告诉了我们 AD 的长度。因为 $ABCD$ 在一条直线上，所以 AD 的长度就等于 $AB+BC+CD$ 的长度。故条件 2 充分。

综上，答案为 B。

例题 8

In the figure shown, the triangle is inscribed in the semicircle. If the length of line segment AB is 8 and the length of line segment BC is 6, what is the length of arc ABC?

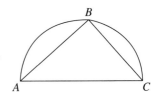

(A) 15π　　　(B) 12π　　　(C) 10π　　　(D) 7π　　　(E) 5π

解：

首先我们要知道，圆的直径对应的内角为 $90°$，即 $\angle B$ 是 $90°$。因此，根据勾股定理，可以得知 $AC = 10$。圆的周长公式为：$\pi d = 10\pi$。arc ABC 是半圆，所以长度为 5π.

答案为 E。

例题 9

The points R, T, and U lie on a circle that has radius 4. If the length of arc RTU is $\dfrac{4\pi}{3}$, what is the length of line segment RU?

(A) $\dfrac{4}{3}$　　　(B) $\dfrac{8}{3}$　　　(C) 3　　　(D) 4　　　(E) 6

解：

依题意，圆的周长为：$2 \times 4 \times \pi = 8\pi$。

已知弧线 RTU 长度为 $\dfrac{4}{3}\pi$，所以它占整个圆周的比例为：$\dfrac{4}{3}\pi \div 8\pi = \dfrac{1}{6}$。

圆周角为 $360°$，则该弧对应的圆心角必为：

$$\frac{1}{6} \times 360° = 60°。$$

因此圆心、R 点和 U 点必然组成一个等边三角形，即 $RU = r = 4$。

答案为 D。

4.5 ▸ 长方体和圆柱体

GMAT 对于立体几何的考查十分简单，主要是一些计算表面积、体积等公式的记忆。

长方体（rectangular solid）是一个三维图形，有六个长方形表面（face）。每个实线和虚线均表示一个边（edge），每个点叫作顶点（vertex）。一个长方体有 6 个面、12 条边和 8 个顶点。所有边的长度都相等的特殊长方体叫作立方体（cube）。

长方体的表面积为所有长方形的面积之和，体积等于"长乘宽乘高"，形如下所示。

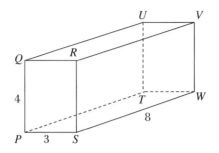

该长方体的表面积为：$2 \times (3 \times 4) + 2 \times (3 \times 8) + 2 \times (4 \times 8) = 136$；体积为：$3 \times 4 \times 8 = 96$。

下图所示是圆柱体（cylinder）。

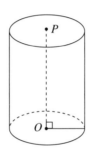

上下两个圆，圆心分别是 O 和 P。高为线段 OP。表面积为两个圆的面积和一个长方形的面积之和，即 $2\pi r^2 + 2\pi rh$；体积为：$\pi r^2 \times h$。

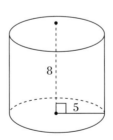

在上图的圆柱体中，表面积为：$2\pi \times 5^2 + 2\pi \times 5 \times 8 = 130\pi$；体积为：$\pi \times 5^2 \times 8 = 200\pi$。

例题 1

For the cube shown, what is the degree measure of $\angle PQR$?

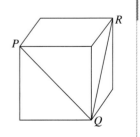

（A）30　　　（B）45　　　（C）60

（D）75　　　（E）90

解：

本题需要作一条辅助线，如右图所示。

连接 PR，因为是立方体，所以 $PR = PQ = QR$。因此三角形 PQR 是等边三角形，任何一个内角均为 $60°$，答案为 C。

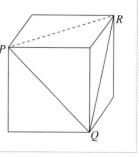

例题 2

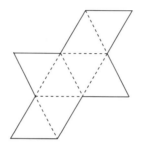

When the figure above is cut along the solid lines, folded along the dashed lines, and taped along the solid lines, the result is a model of a geometric solid. This geometric solid consists of 2 pyramids, each with a square base that they share. What is the sum of the number of edges and the number of faces of this geometric solid?

(A) 10   (B) 18   (C) 20   (D) 24   (E) 25

解：

本题难度较大，考的是空间想象思维能力。比如金字塔的形状，底面是一个四边形，而四边的面是三角形，可以通过空间构想，在脑中折出形状来数有多少个边和面。

不过，本题其实并不需要想象出具体形状。根据题干描述可知，不论折成什么形状，每条虚线单独成为一个 edge，每两条实线重合成一个 edge，总共 7 条虚线和 10 条实线，所以一共会有 12 个 edges；又因为整个图形是一个凸多边形，所以图中每个三角形单独组成一个 face，一共 8 个 faces；所以 edges 加 faces 一共 20 个，答案为 C。

例题 3

What is the volume of the right circular cylinder X?

(1) The height of X is 20.

(2) The base of X has area 25π.

(A) Statement (1) ALONE is sufficient, but statement (2) alone is not sufficient.

(B) Statement (2) ALONE is sufficient, but statement (1) alone is not sufficient.

(C) BOTH statements TOGETHER are sufficient, but NEITHER statement ALONE is sufficient.

(D) EACH statement ALONE is sufficient.

(E) Statements (1) and (2) TOGETHER are NOT sufficient.

解：

题目问的是圆柱的体积。其公式为：$(\pi \times r^2) \times h$。

条件 1 说，高为 20。显然，只知道 h，无法求解，故条件 1 不充分。

条件 2 说，底面积为 25π。只知道底面积，亦不能求解，故条件 2 不充分。

两个条件同时成立时，可以求解，故条件 1 + 条件 2 充分。

综上，答案为 C。

例题 4

In the rectangular solid shown, the three sides shown have areas 12, 15, and 20, respectively. What is the volume of the solid?

(A) 60　　(B) 120　　(C) 450

(D) 1,800　　(E) 3,600

解：

长方体的体积为：长 × 宽 × 高。题干中给出了三个面的面积。如果设 x，y，z 分别为三条边的长，那么这道题会变成一个非常复杂的解方程组问题。实际上，仔细观察可知，三个面的面积其实分别用了长、宽、高两次。若将三个面的面积相乘，则有：

$$(长 \times 宽) \times (高 \times 宽) \times (长 \times 高) = 12 \times 15 \times 20 = 3600,$$

即

$$长^2 \times 宽^2 \times 高^2 = 3600,$$

$$长 \times 宽 \times 高 = 60,$$

答案为 A。

4.6 ▸ 解析几何

解析几何就是借助一个坐标系研究平面直线或几何图形。先来了解直角坐标系的概念。

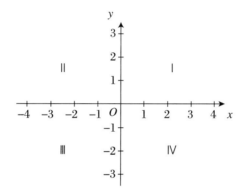

上图就是一个直角坐标系。水平的是 x 轴（x-axis），竖直的是 y 轴（y-axis）。中心的 O 点是原点（origin）。直角坐标系中的象限分布情况也标明在上图中了。

在直角坐标系中的每个点的坐标由 x 轴数值和 y 轴数值共同构成。一个点是用 (x, y) 来表示的。

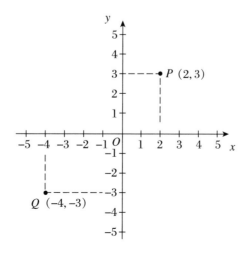

在上图中，点 P 的坐标为（2，3），它处于第一象限。第一象限所有点的坐标均为正；第二象限所有点的坐标为前负后正；第三象限所有点的坐标均为负；第四象限所有点的坐标为前正后负。

直角坐标系上两个点的距离是经常考到的知识点，具体解法如下所述。

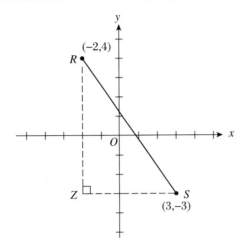

如上图，Z 点的坐标为（–2，–3），$RZ = 7$，$ZS = 5$。因此 RS 的距离为：

$$\sqrt{7^2 + 5^2} = \sqrt{74}。$$

4.6.1 ▶ 直线

直角坐标系中的直线方程为：$y = kx + b$。其中 k 被称为斜率（slope），b 被称为截距（intercept）。下图中的三点都在直线 $y = -\dfrac{1}{2}x + 1$ 上。

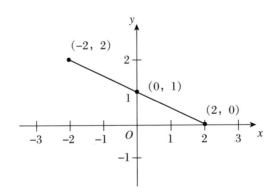

两点可以确定一条直线，这条直线的斜率可以表示为：

$$k = \frac{y_1 - y_2}{x_1 - x_2}$$

当直线斜率为负时，直线向左上方倾斜；当直线斜率为正时，直线向右上方倾斜。

两条直线平行，则斜率相等；两条直线垂直，则斜率相乘等于 -1。

考题中经常出现考查 x 轴截距和 y 轴截距的考题。若问 x 轴截距，只需在直线方程中代入 $y=0$ 求出 x 的值即可；反之，若问 y 轴截距，只需在直线方程中代入 $x=0$ 求出 y 的值即可。

 例题 1

The graph of which of the following equations is a straight line that is parallel to line l in the figure shown?

(A) $3y - 2x = 0$ (B) $3y + 2x = 0$

(C) $3y + 2x = 6$ (D) $2y - 3x = 6$

(E) $2y + 3x = -6$

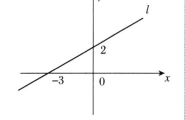

解：

从图中可以看出，直线 l 过点 $(0,2)$ 和点 $(-3,0)$ 可知只有选项 A 满足斜率相等、两直线平行，答案为 A。

 例题 2

In the rectangular coordinate system shown, the area of ΔRST is

(A) $\dfrac{bc}{2}$ (B) $\dfrac{b(c-1)}{2}$

(C) $\dfrac{c(b-1)}{2}$ (D) $\dfrac{a(c-1)}{2}$

(E) $\dfrac{c(a-1)}{2}$

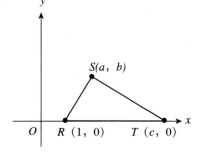

解：

三角形的底边为 R 和 T 的距离，高为 S 的纵坐标。面积为 $\dfrac{(c-1)b}{2}$，答案为 B。

例题 3

In the rectangular coordinate system shown, points O, P, and Q represent the sites of three proposed housing developments. If a fire station can be built at any point in the coordinate system, at which point would it be equidistant from all three developments?

(A) (3, 1)　　　　(B) (1, 3)

(C) (3, 2)　　　　(D) (2, 2)

(E) (2, 3)

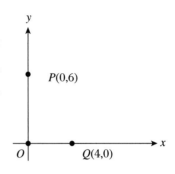

解：

本题有两种方法可以求解。第一种是直接套距离公式，计算比较烦琐，有兴趣的读者可以尝试一下。我们介绍一种比较"直观和直接"的方法。题目要求建造消防站的点到 O，P，Q 三点的距离均相等。OP，OQ 分别在 y 轴和 x 轴上。

首先，我们找到一个到 O，Q 两点距离相等的点，该点必然在 OQ 的中垂线上，因此，所求的点横坐标肯定是 2；同理，到 O，P 两个点距离相同的点在 PO 的中垂线上，所以所求点的纵坐标肯定是 3，因此，所求点的坐标为 (2，3)，答案为 E。

例题 4

In the xy-plane, what is the x-intercept of the line whose equation is $3y - 4x = 15$?

(A) $-\dfrac{15}{4}$　　　(B) $-\dfrac{4}{3}$　　　(C) $\dfrac{4}{3}$　　　(D) $\dfrac{15}{4}$　　　(E) 5

解：

首先将直线方程变换为基本形式，则有：

$$y = \frac{4}{3}x + 5 。$$

注意题干问的是 x 轴的截距，令 $y = 0$，则有 $x = -\dfrac{15}{4}$，答案为 A。

例题 5

Which of the following lines in the xy-plane does not contain any point with integers as both coordinates?

(A) $y = x$ (B) $y = x + \dfrac{1}{2}$ (C) $y = x + 5$

(D) $y = \dfrac{1}{2}x$ (E) $y = \dfrac{1}{2}x + 5$

解：

本题题干有些难懂。其实问的是，选项中哪个直线方程是不含有任何一个 x，y 坐标值均为整数的点。只有选项 B，整理转化为 $y - x = \dfrac{1}{2}$；不存在两个整数的差为分数，两个整数的差只能是整数。

例题 6

In the rectangular coordinate system, line k passes through the point $(n, -1)$. Is the slope of line k greater than zero?

(1) Line k passes through the origin.

(2) Line k passes through the point $(1, n+2)$.

(A) Statement (1) ALONE is sufficient, but statement (2) alone is not sufficient.

(B) Statement (2) ALONE is sufficient, but statement (1) alone is not sufficient.

(C) BOTH statements TOGETHER are sufficient, but NEITHER statement ALONE is sufficient.

(D) EACH statement ALONE is sufficient.

(E) Statements (1) and (2) TOGETHER are NOT sufficient.

解：

题目问的是直线的斜率是否大于 0，也就是 $y = kx + b$ 中的 k 是否大于 0。

条件 1 说，直线通过原点。知道这一点，我们可以将直线方程写为 $y = kx$，代入点 $(n, -1)$，则有：$k = -\dfrac{1}{n}$，即直线方程为：

$$y = -\dfrac{1}{n}x_\circ$$

由于我们不知道 n 的值，所以无法确定斜率是否大于 0，故条件 1 不充分。

条件 2 说，直线通过点 $(1, n+2)$。我们可以将其和点 $(n, -1)$ 一起代入直线方程 $y = kx + b$ 中，

$$n + 2 = k + b,$$
$$-1 = kn + b。$$

第一个式子可以解得：

$$b = n + 2 + k。$$

代入第二个式子中，则有：

$$-1 = kn + n + 2 + k,$$
$$k = \frac{-n-3}{n} + 1。$$

不知道 n 的值，就无法确定 k 的值，故条件 2 不充分。

两个条件同时成立时，可以将点 $(1, n+2)$ 代入 $y = -\frac{1}{n}x$ 中，则有：

$$n + 2 = -\frac{1}{n},$$
$$n^2 + 2n + 1 = 0,$$
$$(n+1)^2 = 0,$$
$$n = -1。$$

能确定 n 的值，必然可以确定 k 的值，故条件 1 + 条件 2 充分。

综上，答案为 C。

例题 7

In the xy-plane, line l and line k intersect at the point $\left(\frac{16}{5}, \frac{12}{5} \right)$. What is the slope of line l?

(1) The product of the slopes of line l and line k is -1.

(2) Line k passes through the origin.

（A）Statement（1）ALONE is sufficient，but statement（2）alone is not sufficient.

（B）Statement（2）ALONE is sufficient，but statement（1）alone is not sufficient.

（C）BOTH statements TOGETHER are sufficient，but NEITHER statement ALONE is sufficient.

（D）EACH statement ALONE is sufficient.

（E）Statements（1）and（2）TOGETHER are NOT sufficient.

解：

条件 1 说，两条直线的斜率相乘为 -1。这个条件表明这两条直线是相互垂直的。但依然无法确定其中某一条的斜率。故条件 1 不充分。

条件 2 说，k 穿过原点。如果知道 k 穿过原点且过点 $\left(\dfrac{16}{5}, \dfrac{12}{5}\right)$，则一定可以知道 k 的方程。但依然无法确定 l 和 k 的关系，进而无法确定 l 的斜率。故条件 2 不充分。

两个条件同时成立时，知道 k 的斜率，且知道 k 和 l 的斜率相乘为 -1，必然可以知道直线 l 的斜率，故条件 1 + 条件 2 充分。

综上，答案为 C。

例题 8

In the rectangular coordinate system shown，if $\triangle OPQ$ and $\triangle QRS$ have equal area，what are the coordinates of point R?

（1）The coordinates of point P are（0，12）.

（2）$OP = OQ$ and $QS = RS$.

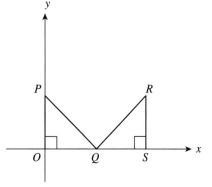

（A）Statement（1）ALONE is sufficient，but statement（2）alone is not sufficient.

（B）Statement（2）ALONE is sufficient，but statement（1）alone is not sufficient.

（C）BOTH statements TOGETHER are sufficient，but NEITHER statement ALONE is sufficient.

（D）EACH statement ALONE is sufficient.

（E）Statements（1）and（2）TOGETHER are NOT sufficient.

解：

条件1，已知点 P 的坐标，由于我们只知道 ΔOPQ 和 ΔQRS 面积相等，所以无法确定点 P 和点 R 的关系，就无法知道点 R 的坐标，故条件1不充分。

条件2，$OP=OQ$，$QS=RS$，整个条件中没有数值，就不可能确定点 R 的坐标，故条件2不充分。

两个条件同时成立时，$OQ=12$，所以 $12 \times 12 = RS \times QS$，又因为 $RS=QS$，所以

$$RS^2 = 144,$$
$$RS = 12 = QS。$$

解得 R 的坐标为（24，12），故条件1+条件2充分。

综上，答案为 C。

例题9

In the rectangular coordinate system shown, does the line k (not shown) intersect quadrant II?

(1) The slope of k is $-\dfrac{1}{6}$.

(2) The y-intercept of k is -6.

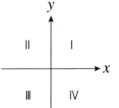

(A) Statement (1) ALONE is sufficient, but statement (2) alone is not sufficient.

(B) Statement (2) ALONE is sufficient, but statement (1) alone is not sufficient.

(C) BOTH statements TOGETHER are sufficient, but NEITHER statement ALONE is sufficient.

(D) EACH statement ALONE is sufficient.

(E) Statements (1) and (2) TOGETHER are NOT sufficient.

解：

题目问的是直线 k 是否经过第二象限。这里我们要知道，斜率是正数的时候，直线必然经过第一象限；斜率是负数的时候，直线必然经过第二象限。因此，想要知道直线是否经过第二象限，知道斜率的正负是关键。

条件1说，k 的斜率是 $-\dfrac{1}{6}$。显然，条件1是充分的。

条件 2 说，k 在 y 轴的截距是 -6。当截距为负数的时候，若直线 k 的斜率是正数，则不会穿过第二象限；若它的斜率是负数，则会穿过第二象限。故条件 2 不充分。

综上，答案为 A。

例题 10

In the rectangular coordinate system，are the points (r, s) and (u, v) equidistant from the origin？

（1）$r + s = 1$.

（2）$u = 1 - r \ and \ v = 1 - s$.

（A）Statement（1）ALONE is sufficient，but statement（2）alone is not sufficient.

（B）Statement（2）ALONE is sufficient，but statement（1）alone is not sufficient.

（C）BOTH statements TOGETHER are sufficient，but NEITHER statement ALONE is sufficient.

（D）EACH statement ALONE is sufficient.

（E）Statements（1）and（2）TOGETHER are NOT sufficient.

解：

题目问的是两点和原点的距离是否相等。根据勾股定理，坐标系上某个点距离原点的距离计算公式为：

$$\sqrt{x^2 + y^2},$$

也就是说，题目问的是 $r^2 + s^2$ 是否等于 $u^2 + v^2$。

条件 1 说，$r + s$ 等于 1；因为 r 和 s 是第一个点的坐标，所以无法判断第二个点（u, v）和原点的距离，自然无法判断两点和原点的距离是否相等，故条件 1 不充分。

条件 2 说，

$$r + u = 1 \ 且 \ s + v = 1,$$

依然无法得出 $r^2 + s^2$ 是否等于 $u^2 + v^2$。故条件 2 不充分。

两个条件同时成立时，可以得出 $u = s$ 且 $r = v$，由此可以得出 $r^2 + s^2 = u^2 + v^2$，故条件 1 + 条件 2 充分。

综上，答案为 C。

4.6.2 ▶ 抛物线

抛物线的标准方程为：$y = ax^2 + bx + c$。

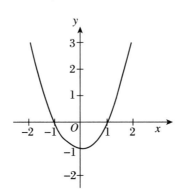

如下几个关于抛物线的知识点是需要牢记的。

（1）二次项常数 a 决定了开口方向。如果 $a > 0$，则开口向上；如果 $a < 0$，则开口向下。

（2）抛物线的对称轴为 $-\dfrac{b}{2a}$。在对称轴处，抛物线会取得极值。

（3）要想判断抛物线是否与 x 轴有交点，需判断 $b^2 - 4ac$ 和 0 的大小关系。

　　若 $b^2 - 4ac > 0$，则抛物线与 x 轴有两个交点；

　　若 $b^2 - 4ac = 0$，则抛物线与 x 轴有一个交点；

　　若 $b^2 - 4ac < 0$，则抛物线与 x 轴没有交点。

例题 11

For what value of x between -4 and 4, inclusive, is the value of $x^2 - 10x + 16$ the greatest?

（A）-4　　　（B）-2　　　（C）0　　　（D）2　　　（E）4

解：

计算抛物线 $y = x^2 - 10x + 16$ 在 $[-4, 4]$ 的值域。根据抛物线的性质可以得出，抛物线开口向上，且对称轴为 $\dfrac{10}{2} = 5$。因此，抛物线在 $[-4, 4]$ 上是单调递减的，即 $x = -4$ 时，y 最大，答案为 A。

例题 12

A certain manufacturer uses the function $C(x) = 0.04x^2 - 8.5x + 25,000$ to calculate the cost, in dollars, of producing x thousand units of its product. The table above gives values of this cost function for values of x between 0 and 50 in increments of 10. For which of the following intervals is the average rate of decrease in cost less than the average rate of decrease in cost for each of the other intervals?

x	$C(x)$
0	25,000
10	24,919
20	24,846
30	24,781
40	24,724
50	24,675

(A) From $x=0$ to $x=10$ (B) From $x=10$ to $x=20$

(C) From $x=20$ to $x=30$ (D) From $x=30$ to $x=40$

(E) From $x=40$ to $x=50$

解:

依题意可知，关于成本的函数是一个开口向上的抛物线函数。对称轴为 $-\dfrac{b}{2a} = 106.25$。仔细观察抛物线的函数图像可知，x 越靠近对称轴时，函数值的变化率一定是越低的。关于这一点，可以想象用一条直线来切抛物线，切点越靠近对称轴，这条直线越接近水平。

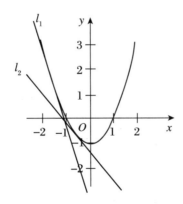

上图中显然直线 l_1 要比直线 l_2 更 "陡"，即变化率更高。

因此，在本题中，x 越靠近 106.25，变化率越低。因此，答案为 E。

几何练习

1. If a square region has area n, what is the length of the diagonal of the square in terms of n?

(A) $\sqrt{2}n$　　(B) $\sqrt{n}$　　(C) $2\sqrt{n}$　　(D) $2n$　　(E) $2n^2$

2. The annual budget of a certain college is to be shown on a circle graph. If the size of each sector of the graph is to be proportional to the amount of the budget it represents, how many degrees of the circle should be used to represent an item that is 15 percent of the budget?

(A) $15°$　　(B) $36°$　　(C) $54°$　　(D) $90°$　　(E) $150°$

3. What is the perimeter, in meters, of a rectangular garden 6 meters wide that has the same area as a rectangular playground 16 meters long and 12 meters wide?

(A) 48　　(B) 56　　(C) 60　　(D) 76　　(E) 192

4. Is the perimeter of square S greater than the perimeter of equilateral triangle T?

(1) The ratio of the length of a side of S to the length of a side of T is 4:5.

(2) The sum of the lengths of a side of S and a side of T is 18.

(A) Statement (1) ALONE is sufficient, but statement (2) alone is not sufficient.

(B) Statement (2) ALONE is sufficient, but statement (1) alone is not sufficient.

(C) BOTH statements TOGETHER are sufficient, but NEITHER statement ALONE is sufficient.

(D) EACH statement ALONE is sufficient.

(E) Statements (1) and (2) TOGETHER are NOT sufficient.

5. In ΔXYZ, what is the length of YZ?

(1) The length of XY is 3.

(2) The length of XZ is 5.

(A) Statement (1) ALONE is sufficient, but statement (2) alone is not sufficient.

(B) Statement (2) ALONE is sufficient, but statement (1) alone is not sufficient.

(C) BOTH statements TOGETHER are sufficient, but NEITHER statement ALONE is sufficient.

(D) EACH statement ALONE is sufficient.

(E) Statements (1) and (2) TOGETHER are NOT sufficient.

6. What is the number of 360-degree rotations that a bicycle wheel made while rolling 100 meters in a straight line without slipping?

(1) The diameter of the bicycle wheel, including the tire, was 0.5 meter.

(2) The wheel made twenty 360-degree rotations per minute.

(A) Statement (1) ALONE is sufficient, but statement (2) alone is not sufficient.

(B) Statement (2) ALONE is sufficient, but statement (1) alone is not sufficient.

(C) BOTH statements TOGETHER are sufficient, but NEITHER statement ALONE is sufficient.

(D) EACH statement ALONE is sufficient.

(E) Statements (1) and (2) TOGETHER are NOT sufficient.

7. What is the radius of the circle above with center O?

(1) The ratio of OP to PQ is 1 to 2.

(2) P is the midpoint of chord AB.

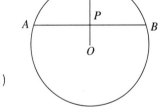

(A) Statement (1) ALONE is sufficient, but statement (2) alone is not sufficient.

(B) Statement (2) ALONE is sufficient, but statement (1) alone is not sufficient.

(C) BOTH statements TOGETHER are sufficient, but NEITHER statement ALONE is sufficient.

(D) EACH statement ALONE is sufficient.

(E) Statements (1) and (2) TOGETHER are NOT sufficient.

8. What is the area of rectangular region R?

(1) Each diagonal of R has length 5.

(2) The perimeter of R is 14.

(A) Statement (1) ALONE is sufficient, but statement (2) alone is not sufficient.

(B) Statement (2) ALONE is sufficient, but statement (1) alone is not sufficient.

(C) BOTH statements TOGETHER are sufficient, but NEITHER statement ALONE is sufficient.

(D) EACH statement ALONE is sufficient.

(E) Statements (1) and (2) TOGETHER are NOT sufficient.

9. In the xy-plane, point (r, s) lies on a circle with center at the origin. What is the value of $r^2 + s^2$?

(1) The circle has radius 2.

(2) The point $(\sqrt{2}, -\sqrt{2})$ lies on the circle.

(A) Statement (1) ALONE is sufficient, but statement (2) alone is not sufficient.

(B) Statement (2) ALONE is sufficient, but statement (1) alone is not sufficient.

(C) BOTH statements TOGETHER are sufficient, but NEITHER statement ALONE is sufficient.

(D) EACH statement ALONE is sufficient.

(E) Statements (1) and (2) TOGETHER are NOT sufficient.

10. In the figure shown, quadrilateral $ABCD$ is inscribed in a circle of radius 5. What is the perimeter of quadrilateral $ABCD$?

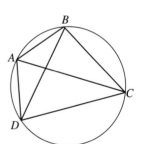

(1) The length of AB is 6 and the length of CD is 8.

(2) AC is a diameter of the circle.

(A) Statement (1) ALONE is sufficient, but statement (2) alone is not sufficient.

(B) Statement (2) ALONE is sufficient, but statement (1) alone is not sufficient.

(C) BOTH statements TOGETHER are sufficient, but NEITHER statement ALONE is sufficient.

(D) EACH statement ALONE is sufficient.

(E) Statements (1) and (2) TOGETHER are NOT sufficient.

11. What is the volume of the cube shown?

(1) The surface area of the cube is 600 square inches.

(2) The length of diagonal AB is $10\sqrt{3}$ inches.

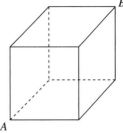

(A) Statement (1) ALONE is sufficient, but statement (2) alone is not sufficient.

(B) Statement (2) ALONE is sufficient, but statement (1) alone is not sufficient.

(C) BOTH statements TOGETHER are sufficient, but NEITHER statement ALONE is sufficient.

(D) EACH statement ALONE is sufficient.

(E) Statements (1) and (2) TOGETHER are NOT sufficient.

12 In the figure shown, lines k and m are parallel to each other. Is $x = z$?

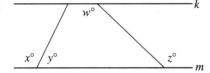

(1) $x = w$

(2) $y = 180 - w$

(A) Statement (1) ALONE is sufficient, but statement (2) alone is not sufficient.

(B) Statement (2) ALONE is sufficient, but statement (1) alone is not sufficient.

(C) BOTH statements TOGETHER are sufficient, but NEITHER statement ALONE is sufficient.

(D) EACH statement ALONE is sufficient.

(E) Statements (1) and (2) TOGETHER are NOT sufficient.

13 In the quilting pattern shown, a small square has its vertices on the sides of a larger square. What is the side length, in centimeters, of the larger square?

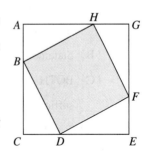

(1) The side length of the smaller square is 10 cm.

(2) Each vertex of the small square cuts 1 side of the larger square into 2 segments with lengths in the ratio of 1:2.

(A) Statement (1) ALONE is sufficient, but statement (2) alone is not sufficient.

(B) Statement (2) ALONE is sufficient, but statement (1) alone is not sufficient.

(C) BOTH statements TOGETHER are sufficient, but NEITHER statement ALONE is sufficient.

(D) EACH statement ALONE is sufficient.

(E) Statements (1) and (2) TOGETHER are NOT sufficient.

14. The circular base of an above-ground swimming pool lies in a level yard and just touches two straight sides of a fence at points A and B, as shown in the figure shown. Point C is on the ground where the two sides of the fence meet. How far from the center of the pool's base is point A?

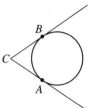

(1) The base has area 250 square feet.

(2) The center of the base is 20 feet from point C.

(A) Statement (1) ALONE is sufficient, but statement (2) alone is not sufficient.

(B) Statement (2) ALONE is sufficient, but statement (1) alone is not sufficient.

(C) BOTH statements TOGETHER are sufficient, but NEITHER statement ALONE is sufficient.

(D) EACH statement ALONE is sufficient.

(E) Statements (1) and (2) TOGETHER are NOT sufficient.

15. A framed picture is shown. The frame, shown shaded, is 6 inches wide and forms a border of uniform width around the picture. What are the dimensions of the viewable portion of the picture?

(1) The area of the shaded region is 24 square inches.

(2) The frame is 8 inches tall.

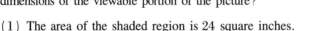
6 in

(A) Statement (1) ALONE is sufficient, but statement (2) alone is not sufficient.

(B) Statement (2) ALONE is sufficient, but statement (1) alone is not sufficient.

(C) BOTH statements TOGETHER are sufficient, but NEITHER statement ALONE is sufficient.

(D) EACH statement ALONE is sufficient.

(E) Statements (1) and (2) TOGETHER are NOT sufficient.

16. In the xy-plane, lines k and l intersect at the point (1, 1). Is the y-intercept of k greater than the y-intercept of l?

(1) The slope of k is less than the slope of l.

(2) The slope of l is positive.

(A) Statement (1) ALONE is sufficient, but statement (2) alone is not sufficient.

(B) Statement (2) ALONE is sufficient, but statement (1) alone is not sufficient.

(C) BOTH statements TOGETHER are sufficient, but NEITHER statement ALONE is sufficient.

(D) EACH statement ALONE is sufficient.

(E) Statements (1) and (2) TOGETHER are NOT sufficient.

17. The perimeters of square region S and rectangular region R are equal. If the sides of R are in the ratio 2: 3, what is the ratio of the area of region R to the area of region S?

(A) 25:16   (B) 24:25   (C) 5:6   (D) 4:5   (E) 4:9

18. Tanks A and B are each in the shape of a right circular cylinder. The interior of Tank A has a height of 10 meters and a circumference of 8 meters, and the interior of Tank B has a height of 8 meters and a circumference of 10 meters. The capacity of Tank A is what percent of the capacity of Tank B?

(A) 75%   (B) 80%   (C) 100%   (D) 120%   (E) 125%

19. In the xy-plane, the distance between the origin and the point (4, 5) is the same as the distance between which of the following two points?

(A) (-3, 2) and (-7, 8)   (B) (-2, 1) and (3, 5)

(C) (-2, -4) and (1, 0)   (D) (3, 2) and (8, 7)

(E) (4, 1) and (-1, -4)

20. What is the volume of the right circular cylinder X?

(1) The height of X is 20.

(2) The base of X has area 25π.

(A) Statement (1) ALONE is sufficient, but statement (2) alone is not sufficient.

(B) Statement (2) ALONE is sufficient, but statement (1) alone is not sufficient.

(C) BOTH statements TOGETHER are sufficient, but NEITHER statement ALONE is sufficient.

(D) EACH statement ALONE is sufficient.

(E) Statements (1) and (2) TOGETHER are NOT sufficient.

几何练习答案及解析

1. A。

略。

2. C。

圆周角是360°，15%的预算占的度数 = 15% ×360 = 54。

3. D。

已知两个长方形面积相等，求其中一个长方形的长 L。$6L = 12 \times 16$，计算得出 $L = 32$，长方形的周长 $= 2 \times (32 + 6) = 76$。

4. A。

条件 1 说，正方形边长和等边三角形边长比为 4:5，所以设三角形边长为 $5a$，正方形边长为 $4a$，其中 a 为大于 0 的数，此时正方形周长为 $16a$，三角形周长为 $15a$，$16a > 15a$，题目可求解。故条件 1 充分。

条件 2 说，正方形边长与等边三角形边长和为 18，但不能确定正方形边长和等边三角形边长的具体值，也不能确定其周长大小关系。故条件 2 不充分。

5. E。

题目并没有给出三角形具体是什么三角形，所以，即使同时给出条件 1 和条件 2，知道三角形的两条边的长度，也无法求得第三条边的长度。故条件 1 + 条件 2 不充分。

6. A。

自行车每转360°，前进的距离为车轮的周长。

条件 1，直径 $d = 0.5$m，周长为 $\pi d \approx 1.57$m，所以前进 100m，旋转的圈数为：$\dfrac{100}{1.57} \approx 63.7$ 圈，题目可求解。故条件 1 充分。

条件 2，转圈的多少只与半径有关，与转速无关，因此条件 2 无法求解。故条件 2 不充分。

7. E。

本题求圆的半径。

若满足条件 1，则有 $PQ = 2OP$，无法求得圆的半径。故条件 1 不充分。

若满足条件 2，则有 $AP = PB$，也无法求得圆的半径。故条件 2 不充分。

结合条件 1 和条件 2，由于无法知道任何一条线段的长度，也无法求得半径。故条件 1 + 条件 2 不充分。

8. C。

设长方形的长为 x，宽为 y，题目求 xy 的值。

条件 1，对角线长为 5，根据勾股定理有：$x^2 + y^2 = 25$，无法确定 xy 的值。故条件 1 不充分。

条件 2，长方形的周长为 14，即 $2(x+y) = 14$，$x + y = 7$，也无法确定 xy 的值。故条件 2 不充分。

结合条件 1 和条件 2，将 $x + y = 7$ 两边同时平方得 $x^2 + 2xy + y^2 = 49$，又因为 $x^2 + y^2 = 25$，所以有 $2xy = 49 - 25 = 24$，即 $xy = 12$。长方形面积为 12。故条件 1 + 条件 2 充分。

9. D。

点 (r, s) 位于以原点为圆心的圆上，有 $r^2 + s^2 = R^2$，其中 R 为圆的半径。

条件 1 给出了圆的半径，所以 $r^2 + s^2 = R^2 = 4$，题目可求解。故条件 1 充分。

条件 2，点 $(\sqrt{2}, -\sqrt{2})$ 也满足 $r^2 + s^2 = R^2$，$(\sqrt{2})^2 + (-\sqrt{2})^2 = R^2$，半径 $R = 2$，题目可求解。故条件 2 充分。

10. C。

条件 1，已知 AB 和 CD 长，无法求得 AD 和 BC 长，也就无法求得四边形 $ABCD$ 的周长。故条件 1 不充分。

条件 2，AC 是直径，所以 $AC = 2 \times 5 = 10$，仅知道 AC 长，也无法计算四边形 $ABCD$ 的周长。故条件 2 不充分。

结合条件1和条件2，由定理"直径所对的圆周角为直角"可知，$\angle ABC$ 和 $\angle ADC$ 均为直角，由勾股定理得：

$$AD = \sqrt{AC^2 - CD^2}。$$

即 $AD = \sqrt{10^2 - 8^2} = 4$，$BC = \sqrt{AC^2 - AB^2}$，即 $BC = \sqrt{10^2 - 6^2} = 8$。

所以，四边形 $ABCD$ 的周长为 $AB + BC + CD + AD = 6 + 8 + 4 + 8 = 26$。故条件1＋条件2充分。

11 ▪▪ D。

求立方体的体积，需要求得立方体的边长，设边长为 x。

条件1，已知立方体的表面积为 600 平方英寸，每个面的面积为 100 平方英寸，则边长为 10 英寸，立方体的体积为 $10^3 = 1000$ 立方英寸。题目可求解，故条件1充分。

条件2，已知对角线 AB 的长为 $10\sqrt{3}$，利用勾股定理，$(x^2 + x^2) + x^2 = (10\sqrt{3})^2$，解得 $x^2 = 100$，$x = 10$，所以立方体的体积为 $10^3 = 1000$ 立方英寸。题目可求解，故条件2充分。

12 ▪▪ D。

条件1，直线 k 和 m 平行，根据定理"两条直线平行，内错角相等"可知 $w = z$，又有 $x = w$，所以 $x = z$，题目可以求解，故条件1充分。

条件2，y 和 x 互为补角，因此 $y = 180 - x = 180 - w$，$x = w$。又因为 $w = z$，所以 $x = z$，题目可以求解。故条件2充分。

13 ▪▪ C。

条件1，已知小正方形的边长为 $DF = 10\text{cm}$，则 $DE^2 + EF^2 = DF^2 = 100$，并不能求得 CE 的长。故条件1不充分。

条件2，点 D 将 CE 切割为 1:2 的两段，即 $DE = 2CD$，又因为 $CD = EF$，所以有 $DE = 2EF$。不能解得 CE 长，故条件2不充分。

综合两个条件，将条件2代入条件1，$4EF^2 + EF^2 = 5EF^2 = 100$，解得 $EF = \sqrt{20} = 2\sqrt{5}$。又由条件2求得，$DE = 4\sqrt{5}$，所以，大正方形边长 $CE = 6\sqrt{5}$。故条件1＋条

件 2 充分。

14 ■ A。

本题求圆的半径。

条件 1 给出圆的面积为 250 平方英尺，所以根据圆的面积公式 $\pi r^2 = 250$ 即可解得 r，即 A 点到圆中心的距离。题目可求解，故条件 1 充分。

条件 2 给出 CQ 的长为 20 英尺，CA 为圆的切线，根据定理 "切线垂直于过切点的半径"，所以 CA 和 QA 垂直，但由条件 2 不能得知 CA 的长，也就无法通过勾股定理求得 QA 的长。故条件 2 不充分。

15 ■ C。

设中间部分的宽为 x，长为 y，因此，阴影部分的宽 $d = \dfrac{6-x}{2}$，因此，本题中要求得中间部分的的面积，必然要求 x 和 y 的值，或者 $x \times y$ 的值，需要两个方程才能满足要求。

条件 1，阴影部分的面积为 24，即 $6 \times (y + 2d) - x \times y = 24$。题目不能求解，故条件 1 不充分。

条件 2，$y + 2d = y + 6 - x = 8$。题目不能求解，故条件 2 不充分。

综合条件 1 和条件 2，将 $y + 2d$ 代入条件 1，即 $6 \times 8 - x \times y = 24$，$x \times y = 24$，中间空白区域的面积为 24。条件 1 + 条件 2 充分。

16 ■ A。

设直线 k 的斜率为 a，直线 l 的斜率为 b，两条直线均过 (1, 1) 点，所以，直线 k 的解析式为：$y = a(x - 1) + 1$，截距为 $1 - a$。直线 l 的解析式为：$y = b(x - 1) + 1$，截距为 $1 - b$。题目要判断是否 $1 - a > 1 - b$，即判断是否 $b > a$。

条件 1，直线 k 的斜率 a 小于直线 l 的斜率 b，能够判断直线 k 的截距大于直线 l 的截距。题目可求解，故条件 1 充分。

条件 2，直线 l 的斜率为正，无法得知直线 k 的斜率和直线 l 的斜率的关系，也就无法判断截距的关系。题目不能求解，故条件 2 不充分。

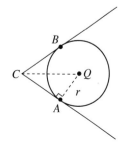

17. B。

设正方形的边长为 x，长方形的宽为 $2k$，长为 $3k$。$4x = 5k$，$x^2 : 6k^2 = 24 : 25$。

18. B。

circumference 是"圆周长"的意思。

19. B。

利用两点间的距离公式可以求解。

20. C。

圆柱体的体积 = 底面积 × 高。

第五章

文字问题

文字问题也被称为"应用题"。顾名思义，应用题是那些用语言或文字叙述有关事实，反映某种数量关系，并求解未知数的题目。在解应用题时，最重要的是进行数学建模，即把文字变成数学表达式。GMAT 数学考试中绝大部分的应用题都不困难，只需仔细读懂题意即可。

例题 1

A cash register in a certain clothing store is the same distance from two dressing rooms in the store. If the distance between the two dressing rooms is 16 feet, which of the following could be the distance between the cash register and either dressing room?

Ⅰ. 6 feet　　　Ⅱ. 12 feet　　　Ⅲ. 24 feet

（A）Ⅰ only　　　（B）Ⅱ only　　　（C）Ⅲ only

（D）Ⅰ and Ⅱ　　　（E）Ⅱ and Ⅲ

解：

这道题其实是一个几何问题。首先，两个试衣间相距 16 feet，要想收款台到两个试衣间距离相等，则收款台需要在两点连线的中轴线上。

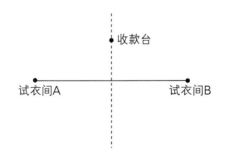

因此，距离最短为 8。也就是说，只要距离大于 8 都是可能的，小于 8 是不可能的。

综上，答案为 E。

例题 2

Of the books standing in a row on a shelf, an atlas is the 30th book from the left and the 33rd book from the right. If 2 books to the left of the atlas and 4 books to the right of the atlas are removed from the shelf, how many books will be left on the shelf?

(A) 56   (B) 57   (C) 58   (D) 61   (E) 63

解:

地图册是书架上从左边数第 30 本书，这表示它的左边有 29 本书；从右边数第 33 本书，这表示它的右边有 32 本书。

因此，书架上原有 $29 + 1 + 32 = 62$ 本书。

一共拿走了 6 本，所以剩余书的数量是

$$62 - 6 = 56。$$

答案为 A。

例题 3

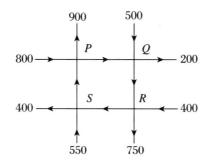

The figure above represents a network of one-way streets. The arrows indicate the direction of traffic flow and the numbers indicate the amount of traffic flow into or out of each of the four intersections during a certain hour. During that hour, what was the amount of traffic flow along the street from R to S if the total amount of traffic flow into P was 1, 200?

(Assume that none of the traffic originates or terminates in the network.)

(A) 200   (B) 250   (C) 300   (D) 350   (E) 400

解：

首先要看懂题目的意思。例如 P 点，它的总流入量 $= 800 + SP$（箭头的方向表示了流入和流出的情况）。每个点流入和流出的总量必须一致，因为不可能有车辆只进入 P 点但不出 P 点。

设 RS 为 x，则有：

$$1200 = 800 + SP,$$

则

$$SP = 400,$$

$$x + 550 = 400 + SP,$$

$$x = 250。$$

答案为 B。

例题 4

A certain brand of house paint must be purchased either in quarts at ＄12 each or in gallons at ＄18 each. A painter needs a 3-gallon mixture of the paint consisting of 3 parts blue and 2 parts white. What is the least amount of money needed to purchase sufficient quantities of the two colors to make the mixture?

（4 quarts ＝1 gallon）

（A）＄54　　　（B）＄60　　　（C）＄66　　　（D）＄90　　　（E）＄144

解：

依题意，首先计算出需要多少蓝色和白色油漆。3 加仑混合油漆按照 3:2 的比例分配，则有

蓝色油漆 $= 3 \times \dfrac{3}{5} = 1.8$。

白色油漆 $= 3 \times \dfrac{2}{5} = 1.2$。

对于蓝色油漆来说：

单独按照夸脱买一定是亏的，直接买 2 加仑的价格为：$18 \times 2 = 36$。

对于白色油漆来说：

买 2 加仑的价格为：$2 \times 18 = 36$，

买 1 加仑和 1 夸脱的价格为：$18 + 12 = 30$。

题目问的是最省钱的模式，则有：

$$36 + 30 = 66。$$

综上，答案为 C。

例题 5

A certain truck traveling at 55 miles per hour gets 4.5 miles per gallon of diesel fuel consumed. Traveling at 60 miles per hour, the truck gets only 3.5 miles per gallon. On a 500-mile trip, if the truck used a total of 120 gallons of diesel fuel and traveled part of the trip at 55 miles per hour and the rest at 60 miles per hour, how many miles did it travel at 55 miles per hour?

(A) 140   (B) 200   (C) 250   (D) 300   (E) 360

解：

依题意，设卡车在 55mile/h 的速度下行驶了 x 英里，则它在 60mile/h 的速度下应行驶了 $500 - x$ 英里。

$$\frac{x}{4.5} + \frac{500-x}{3.5} = 120，$$

解得

$$x = 360。$$

综上，答案为 E。

例题 6

Joanna bought only ＄0.15 stamps and ＄0.29 stamps. How many ＄0.15 stamps did she buy？

（1）She bought ＄4.40 worth of stamps.

（2）She bought an equal number of ＄0.15 stamps and ＄0.29 stamps.

（A）Statement（1）ALONE is sufficient，but statement（2）alone is not sufficient.

（B）Statement（2）ALONE is sufficient，but statement（1）alone is not sufficient.

（C）BOTH statements TOGETHER are sufficient，but NEITHER statement ALONE is sufficient.

（D）EACH statement ALONE is sufficient.

（E）Statements（1）and（2）TOGETHER are NOT sufficient.

解：

题目问的是价值 0.15 美元的邮票买了多少张。

条件 1 说，两种邮票的价值总和为 4.4。则有：$0.15x + 0.29y = 4.4$，且 x 和 y 为整数。

$15x + 29y = 440$。因为 $15x$ 的末尾一定是 5 或 0，所以 $29y$ 的末尾也只能是 0 或 5，y 一定是 5 的倍数。

当 $y = 0$ 时，$x = \dfrac{88}{3}$，不是整数；

当 $y = 5$ 时，$x = \dfrac{59}{3}$，不是整数；

当 $y = 10$ 时，$x = 10$，正确；

当 $y = 15$ 时，$x = \dfrac{1}{3}$，不是整数。

综上，当 $0.15x + 0.29y = 4.4$ 且 x 和 y 为整数时，x 和 y 只能求出一种解，故条件 1 充分。

条件 2 说，两种邮票的数量相等，显然不能得到购买邮票的具体数量，故条件 2 不充分。

综上，答案为 A。

例题 7

The 9 participants in a race were divided into 3 teams with 3 runners on each team. A team was awarded 6-*n* points if one of its runners finished in *n*th place, where $1 \leqslant n \leqslant 5$. If all of the runners finished the race and if there were no ties, was each team awarded at least 1 point?

(1) No team was awarded more than a total of 6 points.

(2) No pair of teammates finished in consecutive places among the top five places.

(A) Statement (1) ALONE is sufficient, but statement (2) alone is not sufficient.

(B) Statement (2) ALONE is sufficient, but statement (1) alone is not sufficient.

(C) BOTH statements TOGETHER are sufficient, but NEITHER statement ALONE is sufficient.

(D) EACH statement ALONE is sufficient.

(E) Statements (1) and (2) TOGETHER are NOT sufficient.

解:

条件 1，没有一组人可以超过 6 分。我们可以直接取极值，假设其中两组都得了最高分 6 分，因此两组总分最高为 12。因为 $1 \leqslant n \leqslant 5$，所以系统的总分为 $1 + 2 + 3 + 4 + 5 = 15$。最后一组至少会有 3 分。故条件 1 充分。

条件 2，没有任意一组名次是连续的。如果第一组获得第 1/3/5 名，第二组获得第 2/4/6 名，则此时第三组 0 分。如果第一组获得第 1/3/6 名，第二组获得第 2/7/8 名，则此时第三组必然获得 3 分。因此条件 2 不充分。

综上，答案为 A。

5.1 ▸ 速率问题

这个问题绝对是小学时期最常见且较容易的问题之一。只需牢记一个公式：

$$速度 = \frac{路程}{时间}。$$

或同类公式:

$$速率 = \frac{总量}{时间}。$$

请注意题干中各变量的单位。

另外，工程问题经常会涉及两台或 n 台机器协作，除非题目特殊注明，否则协作其实就等于各自单独干的简单相加。

例题 1

A certain machine produces 1,000 units of Product P per hour. Working continuously at this constant rate, this machine will produce how many units of Product P in 7 days?

(A) 7,000　　　(B) 24,000　　　(C) 40,000

(D) 100,000　　(E) 168,000

解:
每天有 24h，一天能生产 $1000 \times 24 = 24000$ 件产品，7 天可以生产 $7 \times 24000 = 168000$ 件产品。答案为 E。

例题 2

Machines X and Y run at different constant rates, and Machine X can complete a certain job in 9 hours. Machine X worked on the job alone for the first 3 hours and the two machines working together, then completed the job in 4 more hours. How many hours would it have taken Machine Y, working alone, to complete the entire job?

(A) 18　　　(B) $13\frac{1}{2}$　　　(C) $7\frac{1}{5}$　　　(D) $4\frac{1}{2}$　　　(E) $3\frac{2}{3}$

解:
设工作总量为 1，Y 需要 t 小时独自完成这项工作。

X 单独工作的速率为 $\frac{1}{9}$，Y 单独工作的速率为 $\frac{1}{t}$。

因为 X 单独工作了 3 小时，X 和 Y 又一起工作了 4 小时，所以 X 相当于单独工作了 7 小时，Y 单独工作了 4 小时。

$7 \times \dfrac{1}{9} + 4 \times \dfrac{1}{t} = 1$；$t = 18$，答案为 A。

例题 3

Susan drove at an average speed of 30 miles per hour for the first 30 miles of a trip and then at an average speed of 60 miles per hour for the remaining 30 miles of the trip. If she made no stops during the trip, what was Susan's average speed, in miles per hour, for the entire trip?

(A) 35   (B) 40   (C) 45   (D) 50   (E) 55

解：

前半部分路程是 30 英里，速度是 30 英里每小时，所以时间为 1 小时；后半部分路程是 30 英里，速度为 60 英里每小时，所以时间是 0.5 小时。平均速度为 $\dfrac{总路程}{总时间}$，

即 $\dfrac{60}{1.5} = 40$ 英里/小时，答案为 B。

5.2 ▶ 混合问题

顾名思义，混合问题就是把两种或多种浓度不同的溶液混合到一起，之后再求新溶液浓度的问题。这类问题，一定要把握住一条规则：

混合前后，核心物质的量是不变的。

只要我们能抓住这个物质的量，新溶液的浓度就很容易求得。

例题 1

According to the directions on a can of frozen orange juice concentrate, 1 can of concentrate is to be mixed with 3 cans of water to make orange juice. How many 12-ounce cans of the concentrate are required to prepare 200 6-ounce servings of orange juice?

（A）25　　　　（B）34　　　　（C）50　　　　（D）67　　　　（E）100

解：

一份橘子水中有 $\frac{1}{4}$ 是浓缩液，所以 1 盎司的浓缩粉可以做 4 盎司的橘子汁，总共需要 $200 \times 6 = 1200$ 盎司橘子汁，所以 1200 除以 4 等于 300 盎司浓缩液。因为一罐浓缩液是 12 盎司，所以需要 25 罐浓缩液，答案为 A。

例题 2

Three grades of milk are 1 percent, 2 percent, and 3 percent fat by volume. If x gallons of the 1 percent grade, y gallons of the 2 percent grade, and z gallons of the 3 percent grade are mixed to give $x + y + z$ gallons of a 1.5 percent grade, what is x in terms of y and z?

（A）$y + 3z$　　　　（B）$\dfrac{y + z}{4}$　　　　（C）$2y + 3z$

（D）$3y + z$　　　　（E）$3y + 4.5z$

解：

本题很简单，依照题意列方程为：

$$x + 2y + 3z = 1.5 \times (x + y + z)，$$

解得 $x = y + 3z$，答案为 A。

5.3 ▸ 折扣、利润、税率和利率问题

利润问题和折扣问题的原理一致。

利润问题的公式为：利润 = 收入 × 利率。

折扣问题的公式为：现价 = 原价 × 折扣率。

利率问题稍显复杂，可以分为单利和复利两种情况。

单利：本息和＝本金＋本金×利率×存期

复利：本息和＝本金×（1＋利率）^{存期}

我们需要看清楚题设中给出的计息方式，年复利还是月复利。假设初始金额是 x，年利率是 $r\%$，若按年复利（compounded annually）计息，则一年后的本息和为：

$$(1 + r\%)x。$$

若按月复利（compounded monthly）计息，则一年后的本息和为（年利率＝月利率×12）：

$$\left(1 + \frac{r\%}{12}\right)^{12} x。$$

例题 1

At what simple annual interest rate must \$2,500 be invested if it is to earn \$225 in interest in one year?

（A）6%　　　（B）7%　　　（C）8%　　　（D）9%　　　（E）10%

解：

设利率为 r，则有：

$$2500 \times r = 225,$$

解得 $r = 0.09$，答案为 D。

例题 2

Nathan took out a student loan for \$1,200 at 10 percent annual interest, compounded annually. If he did not repay any of the loan or interest during the first 3 years, which of the following is closest to the amount of interest that he owed for the 3 years?

（A）\$360　　　（B）\$390　　　（C）\$400　　　（D）\$410　　　（E）\$420

解：

看清楚题目，本题是年复利运算。

第一年的本息和为：$1200 + 1200 \times 0.1 = 1320$，

第二年的本息和为：$1320 + 1320 \times 0.1 = 1452$，

第三年的本息和为：$1452 + 1452 \times 0.1 = 1597.2$，

最后一年的本息和减去最开始的本金得397.2，和400最接近，答案为C。

例题 3

If Mel saved more than $10 by purchasing a sweater at a 15 percent discount, what is the smallest amount the original price of the sweater could be, to the nearest dollar?

（A）45　　　（B）67　　　（C）75　　　（D）83　　　（E）150

解：

设 Mel 衣服的原价是 x，则节省的钱为 $15\% \cdot x$。

依据题意，节省的钱要多于 $10，即 $0.15x > 10$，即 $x > 66.7$。答案为B。

例题 3

Judy bought a quantity of pens in packages of 5 for $0.80 per package. She sold all of the pens in packages of 3 for $0.60 per package. If Judy's profit from the pens was $8.00, how many pens did she buy and sell?

（A）40　　　（B）80　　　（C）100　　　（D）200　　　（E）400

解：

设一共买了 x 支笔，则笔的总成本为：

$$\frac{x}{5} \times 0.8。$$

总销售额为：

$$\frac{x}{3} \times 0.6。$$

利润为:

$$\frac{x}{3} \times 0.6 - \frac{x}{5} \times 0.8。$$

题干中利润总额为 8,所以解方程:

$$\frac{x}{3} \times 0.6 - \frac{x}{5} \times 0.8 = 8,$$

解得 $x = 200$,答案为 D。

例题 4

Pat invested x dollars in a fund that paid 8 percent annual interest, compounded annually. Which of the following represents the value, in dollars, of Pat's investment plus interest at the end of 5 years?

(A) $5 \times (0.08x)$　　　　(B) $5 \times (1.08x)$　　　　(C) $[1 + 5 \times 0.08]x$

(D) $(1.08)^5 x$　　　　(E) $(1.08x)^5$

解:

Pat 选择的是年复利的计息方式,所以 5 年后的本息和为:

$$x(1 + 0.08)^5。$$

因此,答案为 D。

例题 5

A merchant purchased a jacket for $60 and then determined a selling price that equaled the purchase price of the jacket plus a markup that was 25 percent of the selling price. During a sale, the merchant discounted the selling price by 20 percent and sold the jacket. What was the merchant's gross profit on this sale?

(A) $0　　　(B) $3　　　(C) $4　　　(D) $12　　　(E) $15

解:

设售卖价为 x,则有:

$$x = 60 + 0.25x,$$

解得
$$x = 80 \text{。}$$

由于折扣为 20%，因此最终售卖价为 64。

$$利润 = 64 - 60 = 4 \text{。}$$

答案为 C。

例题 6

A store bought 5 dozen lamps at ＄30 per dozen and sold them all at ＄15 per lamp. The profit on each lamp was what percent of its selling price?

（A）20%　　（B）50%　　（C）$83\frac{1}{3}$%　　（D）100%　　（E）500%

解：

每打（dozen）是 12 盏台灯。每打台灯为 30 美元，那么每盏台灯的成本为：

$$\frac{30}{12} = 2.5 \text{。}$$

卖出价为每盏台灯 15 美元，则利润为：

$$15 - 2.5 = 12.5 \text{。}$$

利润占卖出价的比值为：

$$\frac{12.5}{15} = 83.33\% \text{。}$$

答案为 C。

例题 7

State X has a sales tax rate of k percent on all purchases and State Y has a sales tax rate of n percent on all purchases. What is the value of $k - n$?

（1）The sales tax on a ＄15 purchase is 30 cents more in State X than in State Y.

（2）The sales tax rate in State X is 1.4 times the sales tax rate in State Y.

（A）Statement（1）ALONE is sufficient, but statement（2）alone is not sufficient.

（B）Statement（2）ALONE is sufficient, but statement（1）alone is not sufficient.

（C）BOTH statements TOGETHER are sufficient, but NEITHER statement ALONE is sufficient.

（D）EACH statement ALONE is sufficient.

（E）Statements（1）and（2）TOGETHER are NOT sufficient.

解：

看题干，我们需要知道 k 和 n 的值。

条件 1 说，对于 15 块钱的东西，X 州的税比 Y 州多 30 分，即 $15 \times k - 15 \times n = 0.3$。显然，我们可以确定 $k - n$ 的值。故条件 1 充分。

条件 2 说，X 州的税率是 Y 州的税率的 1.4 倍。通过本条件，我们可以知道 k 和 n 的比值，但无法知道 $k - n$ 的数值。故条件 2 不充分。

综上，答案为 A。

例题 8

If a merchant purchased a sofa from a manufacturer for ＄400 and then sold it, what was the selling price of the sofa?

（1）The selling price of the sofa was greater than 140 percent of the purchase price.

（2）The merchant's gross profit from the purchase and sale of the sofa was $\frac{1}{3}$ of the selling price.

（A）Statement（1）ALONE is sufficient, but statement（2）alone is not sufficient.

（B）Statement（2）ALONE is sufficient, but statement（1）alone is not sufficient.

（C）BOTH statements TOGETHER are sufficient, but NEITHER statement ALONE is sufficient.

（D）EACH statement ALONE is sufficient.

（E）Statements（1）and（2）TOGETHER are NOT sufficient.

解:

条件 1 说，沙发的卖价比买价的 140% 还要多。由于我们不知道比 140% 多多少，所以条件 1 不充分。

条件 2 说，商人的毛利是卖价的 $\frac{1}{3}$。设卖价是 x，则有：

$$x - 400 = \frac{1}{3}x。$$

显然，可以解出 x 的值，故条件 2 充分。

综上，答案为 B。

例题 9

A store's selling price of ＄2,240 for a certain computer would yield a profit of 40 percent of the store's cost for the computer. What selling price would yield a profit of 50 percent of the computer's cost?

（A）＄2,400　　　　（B）＄2,464　　　　（C）＄2,650

（D）＄2,732　　　　（E）＄2,800

解:

设成本为 x，则有：

$$1.4x = 2240，$$

$$x = 1600。$$

题目问的是 $1.5x$ 的值，即为 2400。

答案为 A。

例题 10

Ellen can purchase a certain computer at a local store at the price of p dollars and pay a 6 percent sales tax. Alternatively, Ellen can purchase the same computer from a catalog for a total of q dollars, including all taxes and shipping costs. Will it cost more for Ellen to purchase the computer from the local store than from the catalog?

(1) $q - p < 50$

(2) $q = 1,150$

(A) Statement (1) ALONE is sufficient, but statement (2) alone is not sufficient.

(B) Statement (2) ALONE is sufficient, but statement (1) alone is not sufficient.

(C) BOTH statements TOGETHER are sufficient, but NEITHER statement ALONE is sufficient.

(D) EACH statement ALONE is sufficient.

(E) Statements (1) and (2) TOGETHER are NOT sufficient.

解:

Ellen 要买电脑,有两种选择:在线下店铺买需花费 p 元加 6% 的税,在线上购买需总共消费 q 元。问的是这两种购买方式哪个划算?实际上,要想知道这道题目的答案,我们至少需要知道 $1.06p$ 是否大于 q。

单独的两个条件明显无法获得确定答案,我们直接来看两个条件结合在一起的情况。

$$q - p < 50 \text{ 且 } q = 1150,$$

$$1150 - p < 50, \text{ 即 } p > 1100,$$

$$1.06p > 1100 \times 1.06 = 1166,$$

由此可知,$1.06p > q$,故条件 1 + 条件 2 充分。

综上,答案为 C。

Last year the price per share of Stock X increased by k percent and the earnings per share of Stock X increased by m percent, where k is greater than m. By what percent did the ratio of price per share to earnings per share increase, in terms of k and m?

(A) $\dfrac{k}{m}\%$　　　　　(B) $(k-m)\%$　　　　　(C) $\dfrac{100(k-m)}{100+k}\%$

(D) $\dfrac{100(k-m)}{100+m}\%$　　　(E) $\dfrac{100(k-m)}{100+k+m}\%$

解:

这道题最重要的是把问题看懂。题目问的是"原先的价格和挣到的钱的比值"的增长率。设原先的价格为 P;原先挣的钱为 E,则有:

原先的比值为: $\dfrac{P}{E}$;

增长后的比值为: $\dfrac{P\left(1+\dfrac{K}{100}\right)}{E\left(1+\dfrac{M}{100}\right)}=\dfrac{P(100+K)}{E(100+M)}$。

增长率的公式为: $\dfrac{现值-原值}{原值}$,即

$$\dfrac{\dfrac{P(100+K)}{E(100+M)}-\dfrac{P}{E}}{\dfrac{P}{E}}=\dfrac{100+k}{100+m}-1=\dfrac{k-m}{100+m}=\dfrac{100(k-m)}{100+m}\%。$$

答案为 D。

At a certain clothing store, customers who buy 2 shirts pay the regular price for the first shirt and a discounted price for the second shirt. The store makes the same profit from the sale of 2 shirts that it makes from the sale of 1 shirt at the regular price. For a customer who buys 2 shirts, what is the discounted price of the second shirt?

(1) The regular price of each of the 2 shirts the customer buys at the clothing store is $16.

（2）The cost to the clothing store of each of the 2 shirts the customer buys is ＄12.

（A）Statement（1）ALONE is sufficient, but statement（2）alone is not sufficient.

（B）Statement（2）ALONE is sufficient, but statement（1）alone is not sufficient.

（C）BOTH statements TOGETHER are sufficient, but NEITHER statement ALONE is sufficient.

（D）EACH statement ALONE is sufficient.

（E）Statements（1）and（2）TOGETHER are NOT sufficient.

解：

本题最重要的是看懂这一句话：The store makes the same profit from the sale of 2 shirts that it makes from the sale of 1 shirt at the regular price.

这句话的意思是：卖两件衣服的利润和卖一件衣服的利润相同。也就是说，第二件衣服，是完全没有利润的。因此，可以得出，第二件衣服的折后价就是它的成本。因此，条件只需要告诉我们第二件衣服的成本即可。

条件 1 说，两件衣服的常规售卖价都是 16 美元。我们无法从中得知两件衣服的成本，故条件 1 不充分。

条件 2 说，两件衣服的成本是 12 美元。显然，第二件衣服的折后价也必然为 12，故条件 2 充分。

综上，答案为 B。

5.4 排列组合

排列组合是组合数学最基本的概念。所谓排列，就是指从给定个数的元素中取出指定个数的元素进行排序。组合则是指从给定个数的元素中仅仅取出指定个数的元素，不考虑排序。

想学会计算排列组合，我们要先学会阶乘。

一个正整数的阶乘（factorial）是所有小于及等于该数的正整数的积，并且 0 的阶乘为 1。

自然数 n 的阶乘写作 $n!$，即 $n! = 1 \times 2 \times 3 \times \cdots \times (n-1) n$。

排列的定义：从 n 个不同元素中，任取 m（$m \leqslant n$，m 与 n 均为自然数，下同）个不同的元素按照一定的顺序排成一列，叫作从 n 个不同元素中取出 m 个元素的一个排列，计算公式为：

$$\mathrm{A}_n^m = n(n-1)(n-2) \cdots (n-m+1) = \frac{n!}{(n-m)!}。$$

组合的定义：从 n 个不同元素中，任取 m（$m \leqslant n$）个元素并成一组，叫作从 n 个不同元素中取出 m 个元素的一个组合，计算公式为：

$$\mathrm{C}_n^m = \frac{\mathrm{A}_n^m}{m!} = \frac{n!}{m!\,(n-m)!}$$

提到排列组合，除了这两个公式外，很多同学的印象必定是杂乱无章的，一个头两个大。实际上，排列组合只需要把脑子理清楚，同时学会两个基本原理，所有题目都能迎刃而解。

5.4.1 ▸ 加法原理和乘法原理

什么是加法原理呢？让我们先举一例。

假设，小明要从北京到上海，现在有 2 班飞机、3 班火车和 2 班长途汽车，问小明一共有多少种方式可以从北京到上海？

很明显，方式应一共是 $2 + 3 + 2 = 7$ 种。

这就是加法原理。之所以能把它们加在一起，是因为无论哪一种方式都能独立地把题目的最终问题解决，即任意一种方式都可以独立完成从北京到上海的任务。

什么是乘法定理呢？再看一例。

假设小明又要从上海回北京，这次没有直达方式了，都需要先经过南京。假设从上海到南京有 3 班火车，从南京到北京有 2 班飞机，问一共有几种方式可以从上海到北京？

这次的方式一共是 $2 \times 3 = 6$ 种。

这就是乘法原理。之所以能把它们乘在一起，是因为无论哪种方式都只是做了一个步骤，无法独立完成任务，需要相乘。

理解了两个基本原理后，我们就可以理解阶乘的实际意义了。假设，我手里有一个黑盒子，盒子里有 6 个颜色不同小球。这 6 个小球，除了颜色不同之外，没有任何不同。那么请问，如果我随机从盒子里取出一个小球，这个小球有几种可能的颜色呢？

答案是 6 种。

那么，再取出一个小球，有几种可能的颜色呢？

显然，答案是 5 种。

以此类推，全部拿出来后，6 个小球可能的颜色种数分别为 6，5，4，3，2，1。从盒子里一个一个取球的过程，其实是相当于给本来在黑盒里排序混乱的小球进行重新排序。因此，如果问 6 个颜色各异的小球之间总共有多少种顺序，那么答案应该是：

$$6 \times 5 \times 4 \times 3 \times 2 \times 1 = 6!。$$

之所以能乘在一起，是因为每个数字只表示一个球可能的颜色，问题是要求排列 6 个小球，所以每个小球都只相当于做了一个步骤。

因此，n 的阶乘的实际意义可以看作给 n 个小球做全排列。

实际情况可能不要求我们给 6 个小球做全排列，只要求排列其中的两个小球。这时，依照刚才的思路则有：

排列 6 个小球中的两个小球 $= 6 \times 5$。

6×5 可以记作 A_6^2，即在 6 个球中排列两个小球。

显然，如果在 6 个小球中排列 6 个小球，那么应该写作 A_6^6，它的值为 6!。

5.4.2 ▶ 除序法

了解完排列的实际意义后，我们就可以了解组合的实际意义了。假设要从 6 个球中选择两个球，然后拿走它们。请问共有多少种不同的选法？

请注意这个问题和排列的区别。如果是从 6 个小球中排列两个小球，那么这两个小球的前后顺序是必须要计算的。例如，排列为红、蓝和蓝、红必然是不同的排法。但在刚才的问题中，只要求选择两个小球拿走，这是不应该考虑两个小球的顺序的，无论是红、蓝还是蓝、红，都是拿走了这两个球，在刚才的问题中应该算作一种情况。

因此，6×5 这个算法多计算了两个小球的顺序。当我们发现算式中多计算了不需要的顺序时，我们可以把这个多出来的顺序除掉，即

$$\frac{6 \times 5}{2!}。$$

$2!$ 表示两个小球的全排列。6×5 中计算了两个小球的顺序，因为题目要求不考虑它们的顺序，所以将顺序除掉。这种不计算取出的小球间的顺序的方式叫作"组合"，记作：C_6^2。

因此，我们经常说，"排列"和"组合"的区别就在于是否需要计算顺序。

5.4.2.1 重复元素

除序法除了在排列组合中可以使用之外，还有两个情况经常使用。第一种情况就是当排列对象中出现重复元素的时候，需要把本来计算过的一些顺序除掉。

例如，A，B，C，D，E 这五个字母的全排列应该为 $5!$。

若问，A，A，B，C，D 这五个字母一共有多少种顺序？

显然，两个 A 是一样的。$5!$ 表示 5 个不同字母的全排列。本题中，因为两个 A 是一样的，这就表示，它们俩谁先谁后是没有区别的，即不应该考虑两个 A 的顺序。利用除序法的定义可知，这五个字母的顺序应为：

$$\frac{A_5^5}{A_2^2} = 60。$$

由此可知，但凡在排列时出现了重复元素，就应该把重复元素的顺序除掉。

定量推理简介
第一章

算数
第二章

代数
第三章

几何
第四章

文字问题
第五章

例题 1

There are 5 cars to be displayed in 5 parking spaces, with all the cars facing the same direction. Of the 5 cars, 3 are red, 1 is blue, and 1 is yellow. If the cars are identical except for color, how many different display arrangements of the 5 cars are possible?

(A) 20　　　(B) 25　　　(C) 40　　　(D) 60　　　(E) 125

解:

因为红色车之间没有区别,所以需要把红色车的顺序除掉。

$$\frac{A_5^5}{A_3^3} = 20。$$

答案为 A。

例题 2

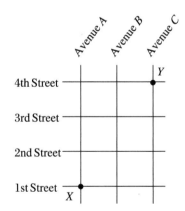

Pat will walk from intersection *X* to intersection *Y* along a route that is confined to the square grid of four streets and three avenues shown in the map above. How many routes from *X* to *Y* can Pat take that have the minimum possible length?

(A) Six　　(B) Eight　　(C) Ten　　(D) Fourteen　　(E) Sixteen

解:

本题很有难度，需要先理解 Pat 行走路线的本质。如果想从 X 走到 Y，最少一定要横着走两格，竖着走三格，题目无非是问我们 Pat 到底什么时候需要横着走，什么时候需要竖着走。我们假设横着走一格是 a，竖着走一格是 b，这个问题就简化成 aabbb 这五个字母的排列问题。例如 aabbb 就是先横着走两格，再竖着走三格，bbbaa 就是先竖着走三格，再横着走两格。aabbb 的排列方式为五个不同字母的全排列除以两个 a 以及三个 b 的顺序：

$$\frac{A_5^5}{A_2^2 A_3^3} = 10。$$

答案为 C。

例题 3

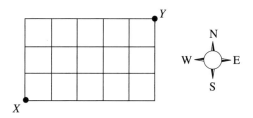

In the figure above, X and Y represent locations in a district of a certain city where the streets form a rectangular grid. In traveling only north or east along the streets from X to Y, how many different paths are possible?

(A) 720 　　　(B) 512 　　　(C) 336 　　　(D) 256 　　　(E) 56

解:

这道题除了表述方式和上一道例题不同外，几乎没有任何区别。把向东走一格假设为 a，向西走一格假设为 b，则我们需要排列 aaaaabbb 的顺序。即

$$\frac{8!}{5! \ 3!} = 56。$$

答案为 E。

5.4.2.2　平均分组问题

除序法的另一个应用是平均分组问题。例如，有 A，B，C，D，E，F 六个人，要求把他们分成两组，一组 4 个人，另一组两个人，请问一共有多少种分法？

这个问题的答案应该是：C_6^2。

即从 6 个人中选出两个人分为一组，剩下的 4 个人自动成为第二组。当然，也可以先选出 4 个人，让剩下的两个人自动成为一组，即 C_6^4。

这两种算法的答案是相等的。

下面再考虑一个问题。如果要求将这 6 个人分为三组，每组两个人，请问有多少种分法。依照刚才的思路，应该是：

$$C_6^2 \cdot C_4^2 \cdot C_2^2。$$

但这个结果是不正确的。为什么呢？试想，假设第一次的挑选情况为：AB CD EF；

第二次的挑选情况为：CD AB EF。

在刚才的算法下，这两次算两种不同的情况。但在实际的问题中，只问有多少种分组方法，并不应该考虑先挑哪组，再挑哪组。也就是说，只要 AB 是一组，CD 是一组，EF 是一组的情况，都应该算作一种情况。因此，当每组人数相同时，应该把多计算的组间顺序除掉，答案应为：

$$\frac{C_6^2 C_4^2 C_2^2}{3!}。$$

此处，3! 表示三组的全排列。

5.4.3 ▶ 桶装信

除了除序法外，排列组合中还有几种经典的模型：

（1）桶装信；

（2）先特殊后一般；

（3）插空、捆绑法；

（4）隔板法；

（5）圆桌问题。

桶装信的最经典版本是：

小明手里有三封信，面前有三个桶。除了信必须要放在桶里外没有任何要求，问一共有几种放法？

这样的问题初看就有两个考虑的方向——"桶"和"信"。

从桶来考虑，这个问题十分复杂：

第一个桶里可能有信可能没信，可能有 1 封、2 封、3 封；第一个桶又会影响到第二个桶。

从信来考虑，这个问题就很简单了：

第一封信，有 3 种可能；第二封信，也有 3 种；第三封信，依然有 3 种。因为最终任务是要放完这 3 封信，所以应该相乘，答案为 $3 \times 3 \times 3 = 9$。

因此，从桶装信问题可以看出，永远都要从信的角度来考虑，千万不要从桶的角度来考虑。类似的背景还有"小孩发糖"和"学生去教室"等。

推广桶装信问题，即所有类似问题都不要从容器角度来出发，要从要放置的东西的角度来出发。答案就是"桶的信次方"。

5.4.4 ▶ 先特殊后一般

很多排列组合的考题条件比较复杂，在计算这类考题时，需要先把有特殊要求的元素安排妥当，再安排其余没什么要求的元素。例如，A，B，C，D，E 五人排成一列，A 要求必须站在第四位上，问一共有几种排法。显然，A 是有特殊要求的那位，所以先把 A 安排在第四位上，只有一种可能性。B，C，D，E 无特殊要求，直接随意安排即可，相当于 4!。答案即为 $1 \times 4! = 24$ 种。

5.4.5 ▶ 插空、捆绑法

当题目要求中提到"某两个元素必须挨在一起"或"某两个元素必须不能挨在一起"时，

需要用到捆绑或插空法。

例如，A，A，C，D，E 五个字母排队，要求 A 和 A 必须在一起，问一共几种排法？

类似这样的问题，我们最先想到的应该是捆绑法，即因为 A 和 A 被要求必须在一起，所以可以将两个 A "捆" 在一起，当成一个元素。这就相当于 4 个元素的排列，即 4!。

插空法和捆绑法的使用情况刚好相反，例如，A，A，C，D，E 五个字母排队，要求两个 A 必须不能挨着，问一共几种排法？

类似这样的问题，我们可以将两个 A 插到 C，D，E 之间，C，D，E 之间有 4 个空位，即 _C_D_E_。插入后再安排 C，D，E 的顺序即可：

$$C_4^2 \cdot A_3^3 = 36。$$

5.4.6 ▶ 隔板法

当给一堆元素分组，且每组元素个数不定时，就会用到隔板法。例如：

小明手里一共有 6 个苹果，要分给 3 个人，要求每个人必须能拿到一个苹果，问有几种分法？

我们可以这样想，将苹果横放一排：

〇　〇　〇　〇　〇　〇

假设手里有两个隔板，把它们插到这六个苹果中间，自然就把这些苹果分成了不定数量的三组。注意，头尾不能放板，因为每个人都必须有一个苹果，如果放在最开头，那么第一个人就分不到苹果了（尾部同理）。

〇｜〇　〇　〇｜〇　〇

如上图这个放法，则第一组 1 个，第二组 3 个，第三组 2 个。因为 6 个苹果中间有五个空位，所以答案为：

$$C_5^2 = 10。$$

5.4.7 ▶ 圆桌问题

普通列成一排时，A，B，C，D，E 五个人排序是 5!。此时，ABCDE 和 BCDEA 是不同的。但当这五个人围坐在一个圆桌上时，开头和结尾变得模糊，ABCDE，BCDEA，CDEAB，DEABC，EABCD 都是相同的。

因为一共五个字母，不论谁开头都是一样的，所以需要把这 5 种顺序除掉，即排序为 $\frac{5!}{5} = 24$ 种。因此，圆桌问题的本质上是除序问题的分支。

若有 n 的元素围成圆桌排序，则一共有 $\frac{A_n^n}{n} = (n-1)!$ 种排序方式。

这些模型分布在排列组合的考题中需要大家自己来判断。大家请记住，无论是什么模型，在排列时我们都要做到不重不漏。

例题 1

Departments A, B, and C have 10 employees each, and Department D has 20 employees. Departments A, B, C, and D have no employees in common. A task force is to be formed by selecting 1 employee from each of departments A, B, and C and 2 employees from Department D. How many different task forces are possible?

(A) 19,000　　(B) 40,000　　(C) 100,000

(D) 190,000　　(E) 400,000

解：
一共 4 组雇员，我们一组一组地选。A 组从 10 人中取 1 个，共有 10 种可能。B，C 两组都和 A 组相同。D 组一共 20 人，选其中 2 个，应为：

$$C_{20}^2 = \frac{A_{20}^2}{A_2^2} = 190。$$

因为最终任务需要分别从 4 组中选择，所以应该全部相乘，即 $10 \times 10 \times 10 \times 190 = 190000$，答案为 D。

例题 2

A certain stock exchange designates each stock with a one-, two-, or three-letter code, where each letter is selected from the 26 letters of the alphabet. If the letters may be repeated and if the same letters used in a different order constitute a different code, how many different stocks is it possible to uniquely designate with these codes?

(A) 2,951　　(B) 8,125　　(C) 15,600

(D) 16,302　　(E) 18,278

解:

本题属于难度相对较大的排列组合考题。首先我们需要先把整个任务分为三种类型，即一个字母、两个字母和三个字母。这三个任务之间是不可能有任何重合的，也不会漏掉任何一种情况。

接下来我们分别看三种情况。因为一共有 26 个字母，所以一个字母的情况非常简单，一共 26 种；因为字母可以重复，因此两个字母的情况为 26×26；三个字母情况同理，为 $26 \times 26 \times 26$。每种情况都能独立完成选码工作，所以总体情况应为三者之和，即 $26 + 26 \times 26 + 26 \times 26 \times 26 = 18,278$，答案为 E。

例题 3

A certain company assigns employees to offices in such a way that some of the offices can be empty and more than one employee can be assigned to an office. In how many ways can the company assign 3 employees to 2 different offices?

(A) 5　　(B) 6　　(C) 7

(D) 8　　(E) 9

解:

该题可以把雇员看作信，办公室看作桶。每个雇员有 2 个选择，答案为 $2 \times 2 \times 2 = 8$。答案为 D。

The letters D, G, I, I, and T can be used to form 5-letter strings as DIGIT or DGIIT. Using these letters, how many 5-letter strings can be formed in which the two occurrences of the letter I are separated by at least one other letter?

(A) 12　　　(B) 18　　　(C) 24　　　(D) 36　　　(E) 48

解:

本题中 the two occurrences of the letter I are separated by at least one other letter 的意思就是两个 I 必然不能相邻的情况，因此应该用插空法。我们可以将两个 I 插到 DGT 之间，DGT 之间有 4 个空位，即 _ D _ G _ T _ 。插入后再安排 DGT 的顺序即可:

$$C_4^2 \cdot A_3^3 = 36。$$

当然，也可以用总数减去两个 I 捆绑在一起这个方法。即

$$A_5^5 - A_4^4 = 96。$$

但你会发现，这个答案是不正确的。为什么？实际上，在计算总体情况时，因为两个 I 一样，所以不能用 5! 来计算，需要把两个 I 的顺序除掉，正确方式应为:

$$\frac{A_5^5}{A_2^2} - A_4^4 = 36。$$

答案是 D。

If x, y, z are both positive integers and $x + y + z = 10$, how many different sets of x, y, z?

(A) 72　　　(B) 54　　　(C) 36　　　(D) 20　　　(E) 18

解：

x, y, z 均不能为 0 且是整数，加在一起和为 10，这就可以看作 10 个苹果分三组，每组必有一个苹果的情况。10 个苹果间有 9 个空位，即

$$C_9^2 = 36。$$

答案为 C。

例题 6

At a dinner party, 5 people are to be seated around a circular table. Two seating arrangements are considered different only when the positions of the people are different relative to each other. What is the total number of different possible seating arrangements for the group?

(A) 5   (B) 10   (C) 24   (D) 32   (E) 120

解：

圆桌问题，排序为 $(5-1)! = 24$ 种方式，答案为 C。

例题 7

In a certain group of 10 members, 4 members teach only French and the rest teach only Spanish or German. If the group is to choose a 3-member committee, which must have at least 1 member who teaches French, how many different committees can be chosen?

(A) 40   (B) 50   (C) 64   (D) 80   (E) 100

解：

题目是问：10 人中有 4 人只教法语，剩下 6 人只教西班语或者德语，从 10 人中选三人组成小队，至少有一个人教法语，有多少种选法？

那么至少有一个教法语的老师有如下这几种情况：

(1) 仅有 1 个法语老师；

（2）有且仅有2个法语老师；

（3）3个都是法语老师。

三种情况的计算分别为：

（1） $C_4^1 C_6^2 = 60$；

（2） $C_4^2 C_6^1 = 36$；

（3） $C_4^3 = 4$。

三者相加为100。

本题还有更简单的解法，那就是所谓的"正难则反"，至少有一个教法语的老师，反面则是一个教法语的老师都没有，在10人中随机选3人的总数减去这个反面的数字，就能得到至少有一个教法语的老师的情况。

$$C_{10}^3 - C_6^3 = 100。$$

答案为 E。

例题 8

From a group of 21 astronauts that includes 12 people with previous experience in space flight, a 3-person crew is to be selected so that exactly 1 person in the crew has previous experience in space flight. How many different crews of this type are possible?

（A）432　　（B）594　　（C）864　　（D）1,330　　（E）7,980

解：

依据题意，可以首先选出那个有经验的宇航员，即 C_{12}^1，

然后从剩下的9个宇航员中选择两个人，即 C_9^2，

两者相乘等于432。答案为 A。

276

定量推理简介
第一章

算数
第二章

代数
第三章

几何
第四章

文字问题
第五章

例题 9

A company plans to assign identification numbers to its employees. Each number is to consist of four different digits from 0 to 9, inclusive, except that the first digit cannot be 0. How many different identification numbers are possible?

(A) 3,024   (B) 4,536   (C) 5,040   (D) 9,000   (E) 10,000

解:

题目是问：某公司要给员工编号，员工的编号由 0~9 中四个不同的数字构成，并且第一位不能为 0，最多有多少个编号？

首先第一位可以选择的数字共有 9 个（除了 0），余下三位将会从 0~9 中剔除了第一位选择的数以后的 9 个数中选三个进行排列，因此共有：

$$9 \times A_9^3 = 4536 \text{。}$$

答案为 B。

例题 10

Ben and Ann are among 7 contestants from which 4 semifinalists are to be selected. Of the different possible selections, how many contain neither Ben nor Ann?

(A) 5   (B) 6   (C) 7   (D) 14   (E) 21

解:

这道题如果充分理解题意，其实非常简单。题目要求的是一定不含有 Ben 和 Ann 的情况，就是要求从剩下的 5 个人里选出 4 个人，答案为：

$$C_5^4 = 5 \text{。}$$

答案为 A。

The letters C, I, R, C, L, and E can be used to form 6-letter strings such as CIRCLE or CCIRLE. Using these letters, how many different 6-letter strings can be formed in which the two occurrences of the letter C are separated by at least one other letter?

(A) 96   (B) 120   (C) 144   (D) 180   (E) 240

解：

题目问的是两个 C 不相邻的情况，因此应该用插空法。我们可以将两个 C 插到 IRLE 之间，IRLE 之间有 5 个空位，即 _ I_ R_ L_ E_ 。插入后再安排 IRLE 的顺序即可：

$$C_5^2 \cdot 4! = 240。$$

答案为 E。

If $n > 4$, what is the value of the integer n?

(1) $\dfrac{n!}{(n-3)!} = \dfrac{3! \; n!}{4! \; (n-4)!}$

(2) $\dfrac{n!}{3! \; (n-3)!} + \dfrac{n!}{4! \; (n-4)!} = \dfrac{(n+1)!}{4! \; (n-3)!}$

(A) Statement (1) ALONE is sufficient, but statement (2) alone is not sufficient.

(B) Statement (2) ALONE is sufficient, but statement (1) alone is not sufficient.

(C) BOTH statements TOGETHER are sufficient, but NEITHER statement ALONE is sufficient.

(D) EACH statement ALONE is sufficient.

(E) Statements (1) and (2) TOGETHER are NOT sufficient.

解：

本题在考查对阶乘算法的理解。

条件 1，我们需要在等式两边同时除以 $\dfrac{n!}{(n-4)!}$，则有：

$$\frac{1}{n-3} = \frac{3!}{4!},$$

显然可以求出 n 的值，故条件 1 充分。

条件 2，等式两边同时除以 $\dfrac{n!}{3!(n-4)!}$，则有：

$$\frac{1}{n-3} + \frac{1}{4} = \frac{n+1}{4 \times (n-3)},$$

整理可得：

$$4 + (n-3) = n+1。$$

此时，无论 n 等于几，两者都相等。因此，无法确定 n 的值，故条件 2 不充分。

综上，答案为 A。

例题 13

A certain university will select 1 of 7 candidates eligible to fill a position in the mathematics department and 2 of 10 candidates eligible to fill 2 identical positions in the computer science department. If none of the candidates is eligible for a position in both departments, how many different sets of 3 candidates are there to fill the 3 positions?

(A) 42   (B) 70   (C) 140   (D) 165   (E) 315

解：

分别从 7 个人中选一个和 10 个人中选两个，则有：

$$C_7^1 \cdot C_{10}^2 = \frac{7 \times 10 \times 9}{2} = 315。$$

答案为 E。

例题 14

Each participant in a certain study was assigned a sequence of 3 different letters from the set {A, B, C, D, E, F, G, H}. If no sequence was assigned to more than one participant and if 36 of the possible sequences were not assigned, what was the number of participants in the study? (Note, for example, that the sequence A, B, C is different from the sequence C, B, A.)

(A) 20　　　(B) 92　　　(C) 300　　　(D) 372　　　(E) 476

解:

这道题的关键在于理解题目。每个人会被分配三个字母, 这三个字母的顺序是需要考虑的, 那么全部的可能性应该是 A_8^3。

有 36 个结果无人认领, 因此参加的总人数应该为:

$$336 - 36 = 300。$$

答案为 C。

例题 15

If a code word is defined to be a sequence of different letters chosen from the 10 letters A, B, C, D, E, F, G, H, I, and J, what is the ratio of the number of 5-letter code words to the number of 4-letter code words?

(A) 5 to 4　　　(B) 3 to 2　　　(C) 2 to 1　　　(D) 5 to 1　　　(E) 6 to 1

解:

选 5 个字母的情况为: A_{10}^5。

选 4 个字母的情况为: A_{10}^4。

两者的比值为:

$$10 \times 9 \times 8 \times 7 \times 6 : 10 \times 9 \times 8 \times 7 = 6 : 1。$$

答案为 D。

例题 16

Two members of a club are to be selected to represent the club at a national meeting. If there are 190 different possible selections of the 2 members, how many members does the club have?

(A) 20　　　　(B) 27　　　　(C) 40　　　　(D) 57　　　　(E) 95

解：

依据题意，设共有 n 个人，则有：

$$C_n^2 = 190,$$

即

$$\frac{n(n-1)}{2} = 190,$$

则，

$$n = 20_\circ$$

综上，答案为 A。

5.5 ▸ 概率论

GMAT 数学考查的概率论主要分为两种——普通概率和独立事件。

5.5.1 ▸ 普通概率

普通概率问题基本属于排列组合的一种应用，只需用"$\dfrac{\text{所求情况}}{\text{总体情况}}$"即可。例如：

一共有 4 颗小球，颜色为红、橙、黄、绿，问抽中红球的概率是多少？

答案是 $\dfrac{1}{4}$，非常简单，抽取一个，一共有 4 种可能，红色只有一种可能，即 $\dfrac{1}{4}$。

需要注意的是依次抽取问题。下面举两个例子。

一共有8颗小球，4红4白，问一次拿出两个球，一红一白的概率是多少？

一共有8颗小球，4红4白，一次只能抽一个，抽取后不放回，问第一次抽到红的，第二次抽到白的概率是多少？

第一个问题的情况非常简单，总体情况为：C_8^2。

抽取两个球，刚好一红一白的可能性为：$C_4^1 \cdot C_4^1$。

因此，概率为：

$$\frac{C_4^1 \cdot C_4^1}{C_8^2} = \frac{4}{7}。$$

第二个问题的情况则需考虑到依次抽取问题。第一次抽取到红色的概率为$\frac{4}{8}$；因为不放回，所以第二次抽取到白球的概率为$\frac{4}{7}$（一个红球已经被抽走了，还剩下7个球），答案为$\frac{4}{8} \times \frac{4}{7} = \frac{16}{56}$。

例题 1

A jar contains 16 marbles, of which 4 are red, 3 are blue, and the rest are yellow. If 2 marbles are to be selected at random from the jar, one at a time without being replaced, what is the probability that the first marble selected will be red and the second marble selected will be blue？

(A) $\frac{3}{64}$　　(B) $\frac{1}{20}$　　(C) $\frac{1}{16}$　　(D) $\frac{1}{12}$　　(E) $\frac{1}{8}$

解：

第一次抽取到红色球的概率为$\frac{4}{16}$；第二次抽取到蓝色球的概率为$\frac{3}{15}$；总共的概率需要将两者相乘，即$\frac{4}{16} \times \frac{3}{15} = \frac{1}{20}$，答案为 B。

例题 2

If a coin has an equal probability of landing heads up or tails up each time it is flipped, what is the probability that the coin will land heads up exactly twice in 3 consecutive flips?

(A) 0.125　　(B) 0.25　　(C) 0.375　　(D) 0.50　　(E) 0.666

解：

硬币抛三次。每次正面朝上和反面朝上的概率相同，均为 $\frac{1}{2}$。连续三次，正面朝上的概率为 $\frac{1}{2} \times \frac{1}{2} \times \frac{1}{2}$。但注意，这还不算结束，问题要求两次正面朝上的情况，因此我们需要选出是哪两次正面朝上，即 C_3^2。

综合概率为：

$$C_3^2 \cdot \frac{1}{2} \cdot \frac{1}{2} \cdot \frac{1}{2} = \frac{3}{8}。$$

答案即选项 C。

当然，我们也可以用传统的做法。首先，每个硬币抛出均有两种可能的结果。一共抛三次硬币，则共有 $2 \times 2 \times 2 = 8$ 种可能。题目要求出现两次正面朝上的情况，则共有三种：第一次和第二次、第一次和第三次、第二次和第三次。

综上，整体概率为 $\frac{3}{8}$。

例题 3

If a certain coin is flipped, the probability that the coin will land heads is $\frac{1}{2}$. If the coin is flipped 5 times, what is the probability that it will land heads up on the first 3 flips and not on the last 2 flips?

(A) $\frac{3}{5}$　　(B) $\frac{1}{2}$　　(C) $\frac{1}{5}$　　(D) $\frac{1}{8}$　　(E) $\frac{1}{32}$

解：

硬币抛三次。每次正面朝上和反面朝上的概率相同，均为 $\frac{1}{2}$。连续五次，概率为 $\frac{1}{2} \times \frac{1}{2} \times \frac{1}{2} \times \frac{1}{2} \times \frac{1}{2}$。问题要求我们前三次正面朝上，后两次反面朝上，所以一共只有一种可能（如果只要求有三次正面朝上，那么就应该从五次里选三次）。因此，综合概率为：

$$\frac{1}{2} \times \frac{1}{2} \times \frac{1}{2} \times \frac{1}{2} \times \frac{1}{2} = \frac{1}{32}。$$

答案为选项 E。

例题 4

When tossed, a certain coin has equal probability of landing on either side. If the coin is tossed 3 times, what is the probability that it will land on the same side each time?

(A) $\frac{1}{8}$　　　(B) $\frac{1}{4}$　　　(C) $\frac{1}{3}$　　　(D) $\frac{3}{8}$　　　(E) $\frac{1}{2}$

解：

硬币抛三次。每次正面朝上和反面朝上的概率相同，均为 $\frac{1}{2}$。连续三次，概率为 $\frac{1}{2} \times \frac{1}{2} \times \frac{1}{2}$。题干要求三次的结果都一样，则共有两种情况——要么都是正面朝上，要么都是反面朝上。因此，综合概率为：

$$2 \times \frac{1}{2} \times \frac{1}{2} \times \frac{1}{2} = \frac{1}{4}。$$

答案为 B。

例题 5

In a stack of cards, 9 cards are blue and the rest are red. If 2 cards are to be chosen at random from the stack without replacement, the probability that the cards chosen will both be blue is $\frac{6}{11}$. What is the number of cards in the stack?

(A) 10　　　　(B) 11　　　　(C) 12　　　　(D) 15　　　　(E) 18

解：

因为卡片不能放回，所以本题是一个重复抽取问题。

设共有 x 张卡片，则

第一次抽取出蓝色卡片的概率为：

$$\frac{9}{x},$$

第二次抽取出蓝色卡片的概率为：

$$\frac{8}{x-1},$$

两次均抽取出蓝色卡片的概率为 $\frac{6}{11}$，两者次概率相乘为 $\frac{6}{11}$，因此，

$$\frac{9}{x} \times \frac{8}{x-1} = \frac{6}{11},$$
$$x(x-1) = 12 \times 11,$$

显然，$x = 12$。

答案为 C。

例题 6

If we have equal chances to go down from A, what is the probability to L?

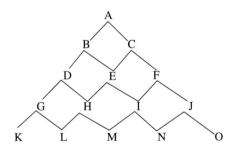

$$(A) \ \frac{1}{2} \qquad (B) \ \frac{1}{4} \qquad (C) \ \frac{1}{16} \qquad (D) \ \frac{1}{32} \qquad (E) \ \frac{1}{64}$$

解：

从 A 到 L，一共要经历 4 层。每层的概率均为 $\frac{1}{2}$。因此，假设从 A 到 L 只有一条

路，则它的概率为 $\frac{1}{2} \times \frac{1}{2} \times \frac{1}{2} \times \frac{1}{2} = \frac{1}{16}$。

实际上，从 A 到 L，不止一种方式，而是有四种方式，分别为 ABDGL，ABDHL，
ABEHL 和 ACEHL。则准确的概率为：

$$4 \times \frac{1}{16} = \frac{1}{4}。$$

综上，答案为 B。

5.5.2 ▸ 独立事件

设 A，B 是两个事件，如果满足等式 $P(AB) = P(A)P(B)$，则称事件 A，B 相互独立，简称 A，B 独立。所谓独立，指的是两个事件之间没有关系，不会因为做了一个而影响到另一个。

独立事件需要记住以下两个公式。

独立事件同时发生的概率为：

$$P(A \cap B) = P(AB) = P(A)P(B)。$$

两个相互独立的事件发生一个事件的概率为：

$$P(A \cup B) = P(A) + P(B) - P(A \cap B)。$$

例题 7

$$M = (-6, \ -5, \ -4, \ -3, \ -2)$$
$$T = (-2, \ -1, \ 0, \ 1, \ 2, \ 3)$$

If an integer is to be randomly selected from set M above and an integer is to be randomly selected from set T above, what is the probability that the product of the two integers will be negative?

(A) 0 (B) $\dfrac{1}{3}$ (C) $\dfrac{2}{53}$ (D) $\dfrac{1}{2}$ (E) $\dfrac{3}{5}$

解：

依题意，从 M 中抽到负数的概率是 100%。如果想让两个数的乘积为负数，则抽取 T 时必须抽到正数。从 T 中抽到正数的概率为 $\dfrac{1}{2}$。由此可知，两个独立事件同时发生的概率为：

$$P(MT) = P(M)P(T) = \dfrac{1}{2}。$$

答案为 D。

例题 8

Let S be a set of outcomes and let A and B be events with outcomes in S. Let $\sim B$ denote the set of all outcomes in S that are not in B and let $P(A)$ denote the probability that event A occurs. What is the value of $P(A)$?

(1) $P(A \cup B) = 0.7$

(2) $P(A \cup \sim B) = 0.9$

(A) Statement (1) ALONE is sufficient, but statement (2) alone is not sufficient.

(B) Statement (2) ALONE is sufficient, but statement (1) alone is not sufficient.

(C) BOTH statements TOGETHER are sufficient, but NEITHER statement ALONE is sufficient.

(D) EACH statement ALONE is sufficient.

(E) Statements (1) and (2) TOGETHER are NOT sufficient.

解：

条件 1，$P(A \cup B) = P(A) + P(B) - P(A \cap B) = 0.7$。由于我们不知道 $P(B)$ 和 $P(A \cap B)$ 的值，所以无法计算 $P(A)$。故条件 1 不充分。

条件 2，$P(A \cup \sim B) = P(A) + [1 - P(B)] - P(A \cap \sim B)$。请注意，$A \cap \sim B$ 表示的是阴影部分：$P(A) + [1 - P(B)] - P(A \cap \sim B) = P(A) + [1 - P(B)] - [P(A) - P(A \cap B)] = 1 - P(B) + P(A \cap B) = 0.9$。由于不知道 $P(B)$ 和 $P(A \cap B)$ 的值，所以无法计算 $P(A)$。故条件 2 不充分。

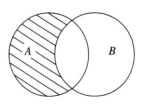

两个条件同时成立时，联立方程组：

$$P(A) + P(B) - P(A \cap B) = 0.7$$
$$-P(B) + P(A \cap B) = -0.1$$

两个方程组相加，则有：

$$P(A) = 0.6。$$

故条件1 + 条件2充分。

综上，答案为C。

例题9

Age	Tax only	Tax and fees	Fees only
18–39	20	30	30
40	10	60	100

The table above shows the number of residents in each of two age groups who support the use of each type of funding for a city initiative. What is the probability that a person randomly selected from among the 250 residents polled is younger than 40, or supports a type of funding that includes a tax, or both?

(A) $\dfrac{1}{5}$ (B) $\dfrac{8}{25}$ (C) $\dfrac{12}{25}$ (D) $\dfrac{3}{5}$ (E) $\dfrac{4}{5}$

解：

从图中可以读出，小于40岁的人数一共有：

$$20 + 30 + 30 = 80。$$

支持包含税的人数一共有：

$$20 + 30 + 10 + 60 = 120。$$

两者的交集有：

$$20 + 30 = 50。$$

由此可知，

$$A \cup B = A + B - A \cap B = 80 + 120 - 50 = 150。$$

选中 $A \cup B$ 的概率为：

$$\frac{150}{250} = \frac{3}{5}。$$

综上，答案为 D。

例题 10

If each of the students in a certain mathematics class is either a junior or a senior, how many students are in the class?

（1）If one student is to be chosen at random from the class to attend a conference, the probability that the student chosen will be a senior is $\frac{4}{7}$.

（2）There are 5 more seniors in the class than juniors.

（A）Statement（1）ALONE is sufficient, but statement（2）alone is not sufficient.

（B）Statement（2）ALONE is sufficient, but statement（1）alone is not sufficient.

（C）BOTH statements TOGETHER are sufficient, but NEITHER statement ALONE is sufficient.

（D）EACH statement ALONE is sufficient.

（E）Statements（1）and（2）TOGETHER are NOT sufficient.

解：

条件 1 说，如果随机抽选一个学生，那么他是 senior 的概率为 $\frac{4}{7}$。从这个条件中，

我们能知道 junior: senior $= 3:4$，但是依然无法确定整体的人数，故条件 1 不充分。

条件 2 说，senior 比 junior 要多 5 个人。这个条件显然无法得出总人数，故条件 2 不充分。

两个条件同时成立时，则有：

$$4x - 3x = 5，$$
$$x = 5。$$

两者的总人数为：$4 \times 5 + 3 \times 5 = 35$。故条件 1 + 条件 2 充分。

综上，答案为 C。

1. A certain bridge is 4,024 feet long. Approximately how many minutes does it take to cross this bridge at a constant speed of 20 miles per hour? (1 mile = 5,280 feet)

(A) 1   (B) 2   (C) 4   (D) 6   (E) 7

2. Five machines at a certain factory operate at the same constant rate. If four of these machines, operating simultaneously, take 30 hours to fill a certain production order, how many <u>fewer</u> hours does it take all five machines, operating simultaneously, to fill the same production order?

(A) 3   (B) 5   (C) 6   (D) 16   (E) 24

3. If Mel saved more than $10 by purchasing a sweater at a 15 percent discount, what is the smallest amount the original price of the sweater could be, to the nearest dollar?

(A) 45   (B) 67   (C) 75   (D) 83   (E) 150

4. One inlet pipe fills an empty tank in 5 hours. A second inlet pipe fills the same tank in 3 hours. If both pipes are used together, how long will it take to fill $\frac{2}{3}$ of the tank?

(A) $\frac{8}{15}$h   (B) $\frac{3}{4}$h   (C) $\frac{5}{4}$h   (D) $\frac{15}{8}$h   (E) $\frac{8}{3}$h

5. A garden center sells a certain grass seed in 5-pound bags at $13.85 per bag, 10-pound bags at $20.43 per bag, and 25-pound bags at $32.25 per bag. If a customer is to buy at least 65 pounds of the grass seed, but no more than 80 pounds, what is the least possible cost of the grass seed that the customer will buy?

(A) $94.03   (B) $96.75   (C) $98.78

(D) $102.07   (E) $105.36

6. Three business partners, Q, R, and S, agree to divide their total profit for a certain year in the ratios 2:5:8, respectively. If Q's share was $4,000, what was the total profit of the business partners for the year?

(A) $26,000   (B) $30,000   (C) $52,000

(D) $60,000   (E) $300,000

7. The price of a certain stock increased by 0.25 of 1 percent on a certain day. By what fraction did the price of the stock increase that day?

(A) $\dfrac{1}{2500}$ (B) $\dfrac{1}{400}$ (C) $\dfrac{1}{40}$ (D) $\dfrac{1}{25}$ (E) $\dfrac{1}{4}$

8. The chart shows year-end values for Darnella's investments. For just the stocks, what was the increase in value from year-end 2000 to year-end 2003?

(A) $1,000   (B) $2,000

(C) $3,000   (D) $4,000

(E) $5,000

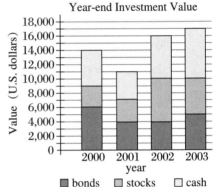

Year-end Investment Value

bonds   stocks   cash

9. At a garage sale, all of the prices of the items sold were different. If the price of a radio sold at the garage sale was both the 15th highest price and the 20th lowest price among the prices of the items sold, how many items were sold at the garage sale?

(A) 33   (B) 34   (C) 35   (D) 36   (E) 37

10. The annual interest rate earned by an investment increased by 10 percent from last year to this year. If the annual interest rate earned by the investment this year was 11 percent, what was the annual interest rate last year?

(A) 1%   (B) 1.1%   (C) 9.1%   (D) 10%   (E) 10.8%

11. Last week Jack worked 70 hours and earned $1,260. If he earned his regular hourly wage for the first 40 hours worked, $\dfrac{3}{2}$ times his regular hourly wage for the next 20 hours worked, and 2 times his regular hourly wage for the remaining 10 hours worked, what was his regular hourly wage?

(A) $7.00 (B) $14.00 (C) $18.00 (D) $22.00 (E) $31.50

12. Last year in a group of 30 businesses, 21 reported a net profit and 15 had investments in foreign markets. How many of the businesses did not report a net profit nor invest in foreign markets last year?

(1) Last year 12 of the 30 businesses reported a net profit and had investments in foreign markets.

(2) Last year 24 of the 30 businesses reported a net profit or invested in foreign markets, or both.

(A) Statement (1) ALONE is sufficient, but statement (2) alone is not sufficient.

(B) Statement (2) ALONE is sufficient, but statement (1) alone is not sufficient.

(C) BOTH statements TOGETHER are sufficient, but NEITHER statement ALONE is sufficient.

(D) EACH statement ALONE is sufficient.

(E) Statements (1) and (2) TOGETHER are NOT sufficient.

13. On Monday morning a certain machine ran continuously at a uniform rate to fill a production order. At what time did it completely fill the order that morning?

(1) The machine began filling the order at 9:30 a. m.

(2) The machine had filled $\frac{1}{2}$ of the order by 10:30 a. m. and $\frac{5}{6}$ of the order by 11:10 a. m.

(A) Statement (1) ALONE is sufficient, but statement (2) alone is not sufficient.

(B) Statement (2) ALONE is sufficient, but statement (1) alone is not sufficient.

(C) BOTH statements TOGETHER are sufficient, but NEITHER statement ALONE is sufficient.

(D) EACH statement ALONE is sufficient.

(E) Statements (1) and (2) TOGETHER are NOT sufficient.

14. Each Type A machine fills 400 cans per minute, each Type B machine fills 600 cans per minute, and each Type C machine installs 2,400 lids per minute. A lid is installed on each can that is filled and on no can that is not filled. For a particular minute, what is the total number of machines working?

(1) A total of 4,800 cans are filled that minute.

(2) For that minute, there are 2 Type B machines working for every Type C machine working.

(A) Statement (1) ALONE is sufficient, but statement (2) alone is not sufficient.

(B) Statement (2) ALONE is sufficient, but statement (1) alone is not sufficient.

(C) BOTH statements TOGETHER are sufficient, but NEITHER statement ALONE is sufficient.

(D) EACH statement ALONE is sufficient.

(E) Statements (1) and (2) TOGETHER are NOT sufficient.

15. Did Insurance Company K have more than $300 million in total net profits last year?

(1) Last year Company K paid out $0.95 in claims for every dollar of premiums collected.

(2) Last year Company K earned a total of $150 million in profits from the investment of accumulated surplus premiums from previous years.

(A) Statement (1) ALONE is sufficient, but statement (2) alone is not sufficient.

(B) Statement (2) ALONE is sufficient, but statement (1) alone is not sufficient.

(C) BOTH statements TOGETHER are sufficient, but NEITHER statement ALONE is sufficient.

(D) EACH statement ALONE is sufficient.

(E) Statements (1) and (2) TOGETHER are NOT sufficient.

16. A tank containing water started to leak. Did the tank contain more than 30 gallons of water when it started to leak? (Note: 1 gallon = 128 ounces)

(1) The water leaked from the tank at a constant rate of 6.4 ounces per minute.

(2) The tank became empty less than 12 hours after it started to leak.

(A) Statement (1) ALONE is sufficient, but statement (2) alone is not sufficient.

(B) Statement (2) ALONE is sufficient, but statement (1) alone is not sufficient.

(C) BOTH statements TOGETHER are sufficient, but NEITHER statement ALONE is sufficient.

(D) EACH statement ALONE is sufficient.

(E) Statements (1) and (2) TOGETHER are NOT sufficient.

17. Each of the 45 books on a shelf is written either in English or in Spanish, and each of the books is either a hardcover book or a paperback. If a book is to be selected at random from the books on the shelf, is the probability less than $\frac{1}{2}$ that the book selected will be a paperback written in Spanish?

(1) Of the books on the shelf, 30 are paperbacks.

(2) Of the books on the shelf, 15 are written in Spanish.

(A) Statement (1) ALONE is sufficient, but statement (2) alone is not sufficient.

(B) Statement (2) ALONE is sufficient, but statement (1) alone is not sufficient.

(C) BOTH statements TOGETHER are sufficient, but NEITHER statement ALONE is sufficient.

(D) EACH statement ALONE is sufficient.

(E) Statements (1) and (2) TOGETHER are NOT sufficient.

18. Stations X and Y are connected by two separate, straight, parallel rail lines that are 250 miles long. Train P and Train Q simultaneously left Station X and Station Y, respectively, and each train traveled to the other's point of departure. The two trains passed each other after traveling for 2 hours. When the two trains passed, which train was nearer to its destination?

(1) At the time when the two trains passed, Train P had averaged a speed of 70 miles per hour.

(2) Train Q averaged a speed of 55 miles per hour for the entire trip.

(A) Statement (1) ALONE is sufficient, but statement (2) alone is not sufficient.

(B) Statement (2) ALONE is sufficient, but statement (1) alone is not sufficient.

(C) BOTH statements TOGETHER are sufficient, but NEITHER statement ALONE is sufficient.

(D) EACH statement ALONE is sufficient.

(E) Statements (1) and (2) TOGETHER are NOT sufficient.

19. If an integer n is to be chosen at random from the integers 1 to 96, inclusive, what is the probability that $n(n+1)(n+2)$ will be divisible by 8?

(A) $\dfrac{1}{4}$ (B) $\dfrac{3}{8}$ (C) $\dfrac{1}{2}$ (D) $\dfrac{5}{8}$ (E) $\dfrac{3}{4}$

20. A certain stock exchange designates each stock with a one-, two-, or three-letter code, where each letter is selected from the 26 letters of the alphabet. If the letters may be repeated and if the same letters used in a different order constitute a different code, how many different stocks is it possible to uniquely designate with these codes?

(A) 2,951   (B) 8,125   (C) 15,600   (D) 16,302   (E) 18,278

定量推理简介
第一章

算数
第二章

代数
第三章

几何
第四章

文字问题
第五章

文字问题练习答案及解析

$1.$ B。

$4024\text{feet} = \dfrac{4024}{5280}\text{mile}$，所求的时间 $t = \dfrac{4024}{5280} \div 20 = \dfrac{4024}{5280 \times 20}\text{h}$，

所求的时间 $\text{minutes} = \dfrac{4024}{5280 \times 20} \times 60 = \dfrac{4024 \times 3}{5280} = \dfrac{4024}{1760} = 2\dfrac{504}{1760} \approx 2.2 \approx 2$。

$2.$ C。

4 台机器同时运行 30 小时完成订单，那么 1 台机器完成这个订单所用的时间就是

$4 \times 30 = 120\text{h}$，如果 5 台机器同时运行，需要的时间是 $\dfrac{120}{5} = 24\text{h}$。注意所求的是 5 台

机器同时运行比 4 台机器同时运行完成订单少用的时间，应该是 $30 - 24 = 6\text{h}$。

$3.$ B

如果 Mel 的衣服原价是 X，则根据题意，$15\% \times X > 10$，整理解得 $X > 66.7$，那么最

小的原价是 67。

$4.$ C。

第一个进水管的 $\dfrac{\text{速度}}{\text{效率}} = \dfrac{1}{5}$，进水 1 个小时注满水池的 $\dfrac{1}{5}$；同理，第二个水管的 $\dfrac{\text{效率}}{\text{速度}} =$

$\dfrac{1}{3}$；两个水管同时开放，注水速度 $= \dfrac{1}{5} + \dfrac{1}{3} = \dfrac{8}{15}$，进水量 $= \dfrac{2}{3}$，时间 $= \dfrac{\frac{2}{3}}{\frac{8}{15}} = \dfrac{5}{4}$。

$5.$ B。

首先计算下三种规格的种子平均每磅的价格。

Items	Cost per pound (平均每磅的价格)
5-pound	$\dfrac{13.85}{5} = 2.77$
10-pound	$\dfrac{20.43}{10} = 2.043$
25-pound	$\dfrac{32.25}{25} = 1.29$

可以看出，容量越大种子平均每磅的价格越便宜，所以如果想花最少的钱，应该尽可能买大包装的种子，就是 25 磅的。顾客买 65 ~ 80 磅的种子，根据题目因为种子的规格只有 5 磅、10 磅、25 磅，都是 5 的倍数，所以顾客买的总重量应该是 5 的倍数，只能是 65 磅、70 磅、75 磅、80 磅。求的是总花费最少，应该尽可能买 25 磅的，然后是 10 磅，最后选择 5 磅。因此，最少的花费应该是 96.75 美元。

6 ▪ B

三个投资人分红的比例是 2:5:8，其中 Q 占的份额是 $\frac{2}{2+5+8} = \frac{2}{15}$，Q 拿到的分红是 4000 美元，可以求出总的利润，$4000 \div \frac{2}{15} = 30000$。

7 ▪ B

本题的关键是要读懂题目。股票增长了 1% 中的 0.25，实际的增长应该是 $0.25 \times 1\% = 0.25\% = \frac{0.25}{100} = \frac{1}{400}$。

8 ▪ B。

从图片中可以看出，2000 年，股票的投资额 $= 9000 - 6000 = 3000$；2003 年，股票的投资额 $= 10000 - 5000 = 5000$；所以可求出 2003 年相对于 2000 年的增长额 $= 5000 - 3000 = 2000$。

9 ▪ B

收音机的价格第 15 高，则有 14 个商品的价格比收音机高；收音机的价格第 20 低，则有 19 个商品的价格比收音机低；商品数量一共就是 $14 + 1 + 19 = 34$。

10 ▪ D

假设去年的投资利率是 L，今年比去年增加了 10%，则今年的投资利率 $= L \times (1 + 10\%) = 11\%$，$L = \frac{11\%}{1 + 10\%} = 10\%$。

11 ▪ B

假设 Jack 的 regular hourly wage 是 x，则 $40x + 1.5x \times 20 + 2x \times 10 = 1260$；$90x = 1260$，$x = 14$。那么 Jack 的基本时薪是 14 美元。

12. D

设有 x 家企业既报告了净利润，也在国外进行投资。不同企业的数量关系如下图所示。左边的圆表示 21 家报告净利润的公司，右边的圆表示 15 家在国外投资的公司，两个圆的相交部分即为既报告了净利润又在国外进行投资的公司。本题求既没有报告净利润又没有在国外投资的公司数量，即求两个圆外的部分。

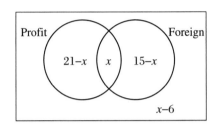

条件 1，30 家公司中有 12 家报告了净利润并在国外进行了投资，即 $x = 12$，那么只报告净利润但没有国外投资的公司数为 $21 - x = 9$，只在国外进行投资没有实现净利润的公司为 $15 - 12 = 3$。所以既没有报告净利润也没有在国外进行投资的公司数为 $30 - 12 - 9 - 3 = 6$。题目可求解，故条件 1 充分。

条件 2，有 24 家企业报告了净利润或投资国外市场，或两者兼而有之。所以既没有报告净利润也没有在国外进行投资的公司数为 $30 - 24 = 6$。题目可求解，故条件 2 充分。

13. B

条件 1，只知道开始时间，不能计算工作时长和结束时间，不满足要求，故条件 1 不充分。

条件 2，从 10:30 到 11:10 的 40 分钟时间里，机器完成了 $\frac{5}{6} - \frac{1}{2} = \frac{1}{3}$ 的工作量，所以每分钟完成 $\frac{1}{3} \times \frac{1}{40} = \frac{1}{120}$ 的工作量，到 11:10 已经完成了 $\frac{5}{6}$ 的工作量，完成剩余的 $\frac{1}{6}$ 工作需要 20 分钟 $\left(\frac{1}{120} \times 20 = \frac{1}{6} \right)$。因此可计算出结束时间为 11:30。题目可求解，故条件 2 充分。

14. C

设 A 类型的机器有 a 台，B 类型的机器有 b 台，C 类型的机器有 c 台。其中 A 和 B

灌装，C 安装盖子。由于题目中说明，所有灌装的罐子均要安装盖子，因此，$400a + 600b = 2400c$，本题求 $a + b + c$ 的值。

条件 1，一分钟总共灌装 4800 罐，$400a + 600b = 4800$，代入得 $2400c = 4800$，解得 $c = 2$。无法得知 a 和 b 的值，所以仅根据条件 1 无法确定 $a + b + c$ 的值，故条件 1 不充分。

条件 2，B 类型的机器是 C 类型的两倍，所以 $b = 2c$。只有条件 2 也无法确定 $a + b + c$ 的值，故条件 2 不充分。

结合条件 1 和条件 2，$c = 2$，$b = 2c = 4$，代入 $400a + 600b = 4800$，解得 $a = 6$，所以所有的机器总数为 $a + b + c = 2 + 4 + 6 = 12$ 台。题目可求解，故条件 1 + 条件 2 充分。

15. E

问公司 K 去年的净利润是否超过 3 亿美元。根据条件 1 和条件 2 可以知道收取保费和支付索赔的利润以及用累计盈余溢价进行投资产生的利润，但不能得知是否有其他收入来源或其他类型的费用。因此，不能计算净利润。

16. E

条件 1，只知道漏水的速率，不知道水漏空的时间，也就不能确定水池开始漏水时装有多少水。故条件 1 不充分。

条件 2，只知道漏水的时间，不知道漏水的速率，也就不能确定水池开始漏水时装有多少水。故条件 2 不充分。

结合条件 1 和条件 2，可知每分钟漏水 6.4 盎司，不到 12 小时，也就是不到 720 分钟就漏空了，判断开始漏水时水池是否装有超过 $30 \times 128 = 3840$ 盎司的水。例如：每分钟漏水 6.4 盎司，漏 100 分钟，满足条件 1 和条件 2，则开始时只有 640 盎司的水，不足 3840 盎司。又如每分钟漏水 6.4 盎司，漏 700 分钟，满足条件 1 和条件 2，则开始时只有 $6.4 \times 700 = 4480$ 盎司的水，超过 3840 盎司。因此，结合条件 1 和条件 2 也无法判断开始漏水时水池是否装有超过 30 加仑的水。

17. B

条件 1，一共有 30 本平装书，因为不知道其中有多少本是西班牙语书，因此也无法判断任取一本书为西班牙语平装书的概率是否小于 $\frac{1}{2}$，故条件 1 不充分。

条件 2，一共有 15 本西班牙书，这 15 本西班牙语书还包括精装和平装，因此，平装西班牙语书的数目小于或等于 15 本，所以任取一本书为西班牙语平装书的概率小于或等于 $\frac{15}{45}=\frac{1}{3}<\frac{1}{2}$，题目可求解，故条件 2 充分。

18■ A

条件 1，相遇时，P 火车的平均时速为每小时 70 英里。所以，相遇时 P 火车行驶的距离为 $70\times2=140$ 英里，此时 P 火车距离终点 $250-140=110$ 英里。所以 Q 火车距离终点 140 英里，P 火车距离终点更近。题目可求解，故条件 1 充分。

条件 2，不知道 P 火车的速度，也就无法计算火车相遇的时间，无法计算相遇时其离终点的距离，故条件 2 不充分。

19■ D

题目的关键是判断三个连续正整数的乘积是否能被 8 整除。

三个连续正整数相乘的组合方式有以下两种：①偶数 × 奇数 × 偶数 ②奇数 × 偶数 × 奇数

第一种情况：相邻两个偶数的乘积一定可以被 8 整除（证明略），所以，所有 n 为偶数的组合都可以被 8 整除。共 48 个偶数。

第二种情况：只有中间偶数为 8 的倍数时，该组合才能被 8 整除。1 ~ 96 中共有 12 个数是 8 的倍数。

所以在 n 取自 1 ~ 96 中的任意整数时，共有 60 种组合可以使 $n(n+1)(n+2)$ 被 8 整除。

则这个概率为 $\frac{60}{96}=\frac{5}{8}$。

20■ E

有三种编码，分别是一位、二位、三位，每一位字母都是 26 个字母表中的任意一个。编码中的字母可以重复，同样的字母如果顺序不同也认为是不同的编码，问一共可以构成多少种编码。

一位编码：只有 26 种。

二位编码：由于顺序不同也是不同的编码，即 ab 和 ba 是两个不同的编码，所以二位编码共有 26^2 种。

同理，三位编码共有 26^3 种。

共有 $26 + 26^2 + 26^3 = 18,278$ 种编码。